Information Systems Engineering and Management

Volume 22

The book series "Information Systems Engineering and Management" (ISEM) publishes innovative and original works in the various areas of planning, development, implementation, and management of information systems and technologies by enterprises, citizens, and society for the improvement of the socio-economic environment.

The series is multidisciplinary, focusing on technological, organizational, and social domains of information systems engineering and management. Manuscripts published in this book series focus on relevant problems and research in the planning, analysis, design, implementation, exploration, and management of all types of information systems and technologies. The series contains monographs, lecture notes, edited volumes, pedagogical and technical books as well as proceedings volumes.

Some topics/keywords to be considered in the ISEM book series are, but not limited to: Information Systems Planning; Information Systems Development; Exploration of Information Systems; Management of Information Systems; Blockchain Technology; Cloud Computing; Artificial Intelligence (AI) and Machine Learning; Big Data Analytics; Multimedia Systems; Computer Networks, Mobility and Pervasive Systems; IT Security, Ethics and Privacy; Cybersecurity; Digital Platforms and Services; Requirements Engineering; Software Engineering; Process and Knowledge Engineering; Security and Privacy Engineering, Autonomous Robotics; Human-Computer Interaction; Marketing and Information; Tourism and Information; Finance and Value; Decisions and Risk; Innovation and Projects; Strategy and People.

Indexed by Google Scholar. All books published in the series are submitted for consideration in the Web of Science.

For book or proceedings proposals please contact Alvaro Rocha (amrrocha@gmail.com).

Shubham Mahajan · Álvaro Rocha ·
Amit Kant Pandit · Paras Chawla
Editors

Smart Systems: Engineering and Managing Information for Future Success

Navigating the Landscape of Intelligent Technologies

Springer

Editors
Shubham Mahajan
Amity School of Engineering
and Technology (ASET)
Amity University
Gurugram, Haryana, India

Amit Kant Pandit
School of ECE
Shri Mata Vaishno Devi University
Katra, Jammu and Kashmir, India

Álvaro Rocha
ISEG
University of Lisbon
Lisbon, Portugal

Paras Chawla
Amity School of Engineering
and Technology (ASET)
Amity University
Gurugram, Haryana, India

ISSN 3004-958X ISSN 3004-9598 (electronic)
Information Systems Engineering and Management
ISBN 978-3-031-76151-5 ISBN 978-3-031-76152-2 (eBook)
https://doi.org/10.1007/978-3-031-76152-2

This Springer imprint is published by the registered company Springer Nature Switzerland AG
The registered company address is: Gewerbestrasse 11, 6330 Cham, Switzerland

Contents

Introduction to Fuzzy Logic and Its Applications in Machine Learning

Ashima Kalra

Abstract This chapter provides an introduction to fuzzy logic and discusses its various applications in engineering and research. Numerous studies are being conducted in the applied branch of mathematics known as fuzzy logic. In Fuzzy Logic, "if…then…" is used. According to fuzzy logic, members are not limited to T or F definitions. Fuzzy logic is particularly used in situations where traditional binary logic is not optimal such as when dealing with uncertain data or on making decisions based on incomplete or conflicting information. It is used in many different areas, including control system, decision making, linguistics, financial analysis, power system, medical diagnosis, gaming, image and signal processing, etc. This also refers to recent advancements or trends in fuzzy logic research and applications that include emerging fields where fuzzy logic is proving particularly promising or ongoing efforts to refine its methodologies.

Keywords Fuzzy logic · Optimization · Fuzzification

1 Introduction

Fuzzy logic is a useful tool for issue resolution that has many applications in information technology, consumer electronics, expert systems, management, and medical fields. It makes it simple to derive certain conclusions from evidence that is imprecise, equivocal, or unclear. This approach to decision-making is innate and closely aligns with human thought processes and decision-making, especially in instances where answers are ambiguous.

As we proceed, we will see that fuzzy logic requires a different way of thinking than a system based on classical logic, which calls for accurate mathematical equations, precise numerical values, and a full understanding of a system. This alternative

A. Kalra (✉)
Department of Electronics and Communication Engineering, Chandigarh Engineering College-CGC, Landran, Punjab, India
e-mail: uppal.ashi@gmail.com

S. Mahajan et al. (eds.), *Smart Systems: Engineering and Managing Information for Future Success*, Information Systems Engineering and Management 22,
https://doi.org/10.1007/978-3-031-76152-2_1

approach models complex systems at a higher level of abstraction grounded in the understanding and expertise of the experts, avoiding the need for the rigorous mathematical treatment of the systems that has up to now dominated the scientific world. Some scientists and engineers still harbor misgivings about fuzzy systems because they believe that a system model devoid of mathematics isn't really a model at all.

We might explain to them that it's a paradigm change, that it is simple to use, that its effectiveness has been proven, and that it operates more like the human mind. Because it allows the articulation of the information required to make a choice using arbitrary concepts like "very hot," "warm," "little cool," etc., fuzzy logic provides an intuitive approach to decision-making.

Comprehending the fundamental workings of a system is essential for managing the most challenging design and development tasks. Thus, this chapter has introduced the concepts and ideas related to the design of fuzzy systems.

2 A Fuzzy System

This section will examine systems based on fuzzy logic (also known as fuzzy inference systems). Figure 1 shows how a block diagram can be used to represent a fuzzy system [1]. Fuzzy system is a type of artificial intelligence based on the concept of fuzzy logic. It aims to address uncertainty and inaccuracies in data and knowledge by using linguistic variables and fuzzy sets instead of traditional binary logic. The components of a fuzzy system include a fuzzifier (which converts crisp input data into fuzzy sets), a knowledge base (which contains rules and membership functions), an inference engine (which applies these rules to make decisions), and a defuzzifier (which converts fuzzy data into fuzzy sets) set). Data converts the output into an unambiguous value (Fig. 2).

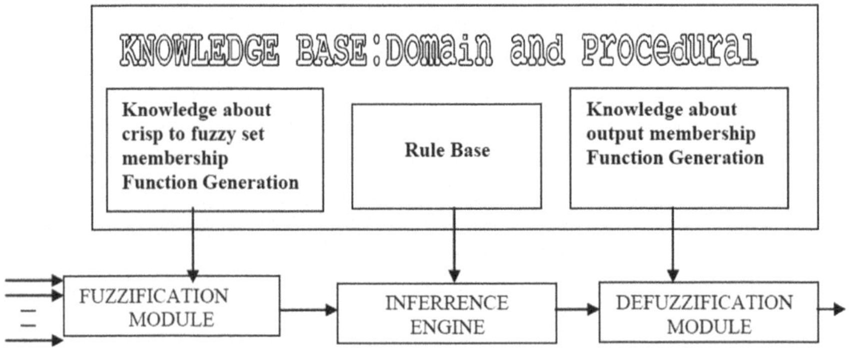

Fig. 1 Fuzzy logic based system

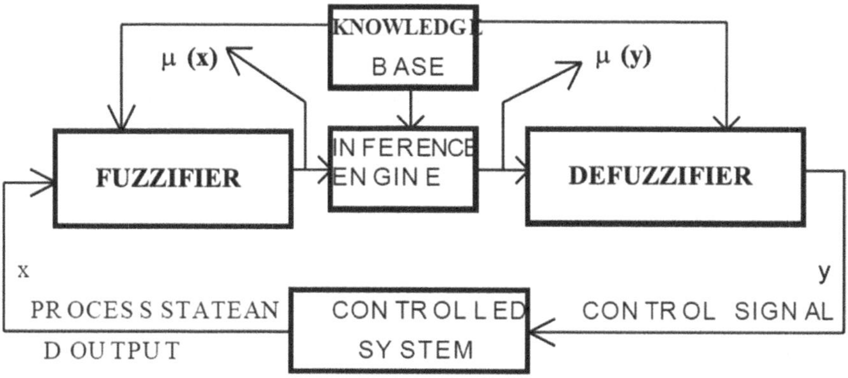

Fig. 2 A fuzzy controller

2.1 *Fuzzification Module*

The fuzzification module is a component of a fuzzy system that transforms crisp input data into fuzzy sets. Its main function is to create membership functions for the linguistic variables that are used in the system. The membership functions ascertain the extent to which an input variable is a member of a fuzzy set.

The fuzzification module maps the input data to these membership functions and determines the extent of truth of each input variable in terms of each fuzzy set. It then generates a fuzzy input vector for the inference module to process and make decisions based on the rules in the knowledge base. The fuzzification module is essential for handling imprecise and uncertain input data, making fuzzy systems suitable for dealing with complex and ambiguous problems.

2.2 *Knowledge Base*

In a fuzzy system, the knowledge base is a component that holds the rules, membership functions, and other information necessary to make decisions based on the input data. It is essentially the "brain" of the system, as it stores the knowledge that the system has acquired through training or other means. The knowledge base is responsible for defining the linguistic variables that the system will use to represent the input and output data, and it also contains the membership functions that define the fuzzy sets associated with those variables. This also has a set of rules that establish relationships among the variables and govern how the system will make decisions based on the input data. The knowledge base can be updated and adjusted as new data becomes available, allowing the system to continually learn and improve its decision-making capabilities.

2.3 *Inference Engine*

In a fuzzy system, the inference engine is the component that performs the reasoning and decision-making based on the fuzzy sets and rules defined in the knowledge base. It takes the fuzzy input data generated by the fuzzification module and applies the rules to derive a fuzzy output. Its objective is to combine the several fuzzy sets to produce a single fuzzy set as the result using the operators specified in the rules. This output represents the system's prediction or decision based on the input data and the rules in the knowledge base. The inference engine can use various methods to perform the inference, such as Mamdani and Takagi–Sugeno methods. It is an important component of a fuzzy system that has a critical impact in transforming fuzzy, uncertain input data into meaningful and actionable decisions.

2.4 *Defuzzification Module*

The following tasks are carried out by the defuzzification module:

1. A scale mapping that creates comparable discourse universes from the output variables' range of values.
2. Defuzzification, which converts hazy results into matching, clear results so that appropriate action can be taken.

3 Working of a Fuzzy System

Figure 3 serves as an illustration of how a fuzzy system operates. Here, we look into a system of compositional inferencing. The following 5 steps are used by a fuzzy system operating within the framework of a compositional inference mechanism to conduct its computation.

Step 1: **Fuzzification**

"Fuzzification" is a method of transforming crisp input data to corresponding data in the fuzzy domain (fuzzy values). At this point, the inputs are applied into the fuzzifier, which employs membership functions to ascertain the relative contributions of each input to each of the pertinent fuzzy sets. The input(s) is always a precise numerical value.

Step 2: **Use fuzzy operators**

Since the antecedent of a given rule typically consists of multiple parts, the fuzzy operators (t-norms and s-norms) are used to combine the results of the antecedents for a given rule into a single number. The output function will then be used with this value. Two or more membership values from fuzzified input variables are the

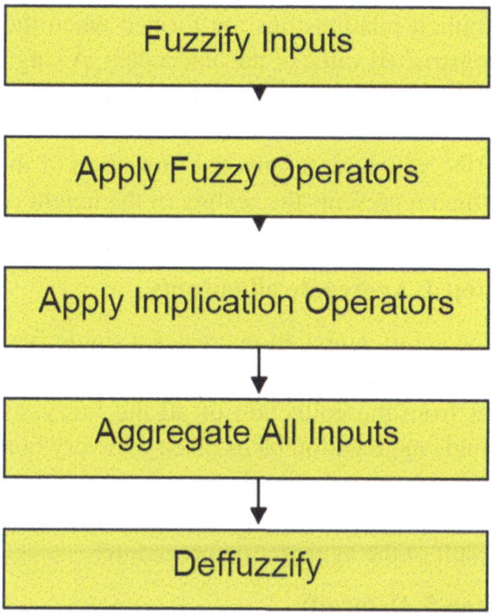

Fig. 3 Fuzzy system

input to a fuzzy operator. One truth value is the result. Here, we simply discuss the two most often used fuzzy operators: "MIN," which stands for "and" connective, and "MAX," which stands for "or" connective. For instance, let's say the following guideline governs how a university department is rated:

IF (EXPLORATION is GOOD) ***and*** *(LEARNER_EMPLOYMENT is HIGH) then (Department_Rating is Excellent)*

And let us further assume that crisp input values of variables EXPLORATION and LEARNER_EMPLOYMENT are such that these have a membership degree of 0.8 and 0.4 in fuzzy set GOOD and HIGH respectively, then **and** operation will result into min (0.8, 0.4) = 0.4

This is graphically represented in Fig. 4

Step 3: **Apply implication methods**

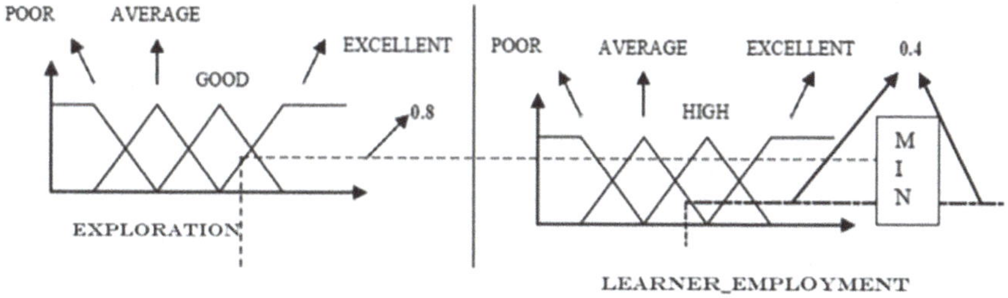

Fig. 4 Composition

Implicit relationships are formed when the consequent is shaped according to the constructed value of the antecedent. A single number received from the composition process serves as the input for the implication process, and the output is a fuzzy set—that is, the result of a fuzzy operator. There are implications for every regulation. MIN, which represents the truncation of the subsequent fuzzy set, and PRODUCT, which represents the scaling of the height of the consequent fuzzy set, are two often employed implication operators.

Step 4: Aggregate all outputs

The act of aggregation creates a single fuzzy set by combining multiple fuzzy sets in a desirable fashion. The process of aggregation involves creating a single fuzzy set from the collection of all the fuzzy sets that each rule's output represents. A single aggregation takes place for every output variable. The list of truncated output functions that the implication process for each rule returns serves as the input of the aggregation process in our model. For every output variable, a single fuzzy set is the result of the aggregation procedure.

Step 5: Defuzzify

The fuzzy values are converted to crisp values using defuzzification [2]. A fuzzy set is the input for the defuzzification process, and a corresponding single crisp integer is the output [3] (Fig. 5).
"light"/>

In fuzzy systems, defuzzification is not a standard procedure. Although there are several defuzzification techniques available, the following are the most often employed techniques:

(a) Centre of Gravity/Centre of Area/Centroid
(b) Centre of Sums
(c) Height defuzzification

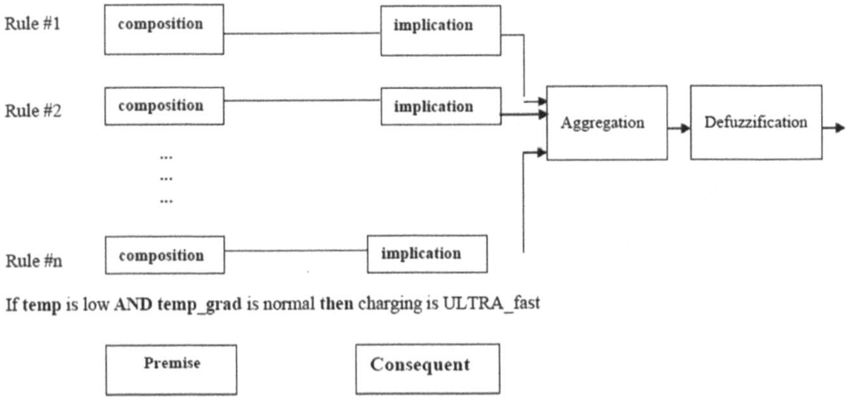

Fig. 5 Composition

(d) Centre of largest area
(e) Min–Max approach
(f) Weighted average Compositional Inference

3.1 *The Inference Process*

The literature has a good number of aggregation operators, a wide range of implications, and a significant number of composition operators. The inference process varies depending on the type of composition operator and implication operator that are utilized. The following are the top three inference mechanisms [4, 5]:

(a) Mamdani-style inference
(b) Larsen-style inference
(c) Sugeno-style inference

Here ***Mamdani-style*** inference is used as it is one of the commonly used fuzzy inference methodology. This approach is expected to provide fuzzy sets as its output membership functions. This method is illustrated in Fig. 6. Every output variable has a single fuzzy set after aggregation, which needs to be defuzzified.

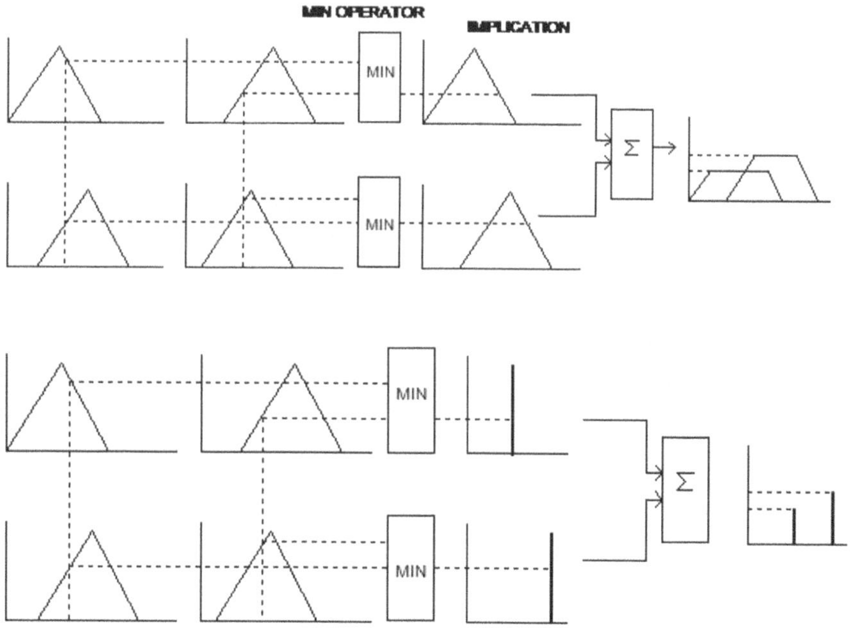

Fig. 6 Composition

4 Fuzzy Logic in Machine Learning

A mathematical framework called fuzzy logic addresses imprecision and ambiguity in data and decision-making. It is particularly useful in machine learning and artificial intelligence for handling situations where data or conditions are not binary (true/false) [6–8] but exist on a continuum. Here are some key aspects of fuzzy logic in machine learning:

Membership Functions: Fuzzy logic uses membership functions to describe the degree of membership of an element in a set. Instead of assigning a binary value of 0 or 1, fuzzy sets allow for values between 0 and 1, representing degrees of truth. Membership functions are used to model linguistic variables (e.g., "hot," "cold," "tall," "short") and their associated uncertainty.

Fuzzy Rules: Fuzzy logic systems consist of a set of fuzzy rules that define how input variables relate to output variables. These rules use linguistic terms and membership functions to express relationships in a more human-readable and interpretable way.

Fuzzy Inference Systems: Fuzzy inference systems (FIS) [6] are used to process fuzzy rules and determine the output based on the input data. FIS typically involve steps such as fuzzification (converting crisp inputs to fuzzy values), rule evaluation (applying fuzzy rules), aggregation (combining rule outputs), and defuzzification (converting fuzzy output to a crisp value).

4.1 Applications: Fuzzy Logic in Machine Learning

Fuzzy logic is a system that can be used in control systems, pattern recognition, natural language processing, and expert systems for the purpose of text analysis. In control systems, fuzzy logic controllers can be implemented for controlling complex systems since accurate mathematical models are unavailable. Some of the application areas include temperature control, speed control, and robotics.

Pattern Recognition: One possible application of fuzzy logic can be pattern recognition and classification problems that have indistinguishable or overlapping classes.

Natural Language Processing: It is another field where Fuzzy Logic can be put into operation with the ease of handling linguistic terms and uncertainty of textual data to facilitate natural language understanding and processing.

In decision support systems, fuzzy logic helps in making decisions where precise or reliable data is unavailable, and is based on vague, unreliable information.

4.2 Advantages: Fuzzy Logic in Machine Learning

Machine learning applications based on fuzzy logic present a number of benefits. Initially, it is the manner in which it mimics the naturalness and intuitiveness of uncertainty and imprecision in data for the ultimate real-world approach; hence permitting more accurate depictions of observed phenomena. Also, systems that employ fuzzy logic are easily interpretable, which consequently can be important when transparency of the decision-making process is an essential aspect, e.g., medical diagnosis or expert system.

Also, fuzzy logic is adept at dealing with nonlinear correlations and intricate decision boundaries effectively; thus, it suits tasks where a multitude of cases are found with uncertain boundaries. Moreover, it has the advantage that it can be used as one element of the same system together with other machine learning methods, making it a hybrid system based on both fuzzy logic and other procedures.

Generally speaking, fuzzy logic is an enhancement of machine learning models with greater flexibility as well as interpretability and can be used effectively as a powerful tool in many domains. It can capture complex nonlinear relationships that might not easily fit within the scope of traditional binary logic. It provides a general approach to handling data where uncertainty and vagueness play a role. Fuzzy logic systems are interpretable, so one can understand and debug the reasoning process more easily.

4.3 Limitatons: Fuzzy Logic in Machine Learning

While fuzzy logic offers valuable advantages, it also has some limitations in machine learning applications. A major disadvantage is that it is difficult to determine appropriate membership functions and rule sets, which often requires expert knowledge in the field and can be very time-consuming. Fuzzy systems can be sensitive to the choice of these parameters, potentially affecting their performance and interpretability. Furthermore, fuzzy logic may not always outperform other machine learning techniques, especially for tasks where deep learning methods produce excellent results, such as: B. Image recognition or natural language processing. The interpretability of fuzzy logic is one of its advantages, but can also be a limitation in situations where a black-box model is preferred for higher accuracy. Overall, while fuzzy logic is a valuable tool, its effectiveness depends on careful design and may not be the best choice for every machine learning problem.

Creating fuzzy rule sets can be a challenging task that demands expertise in the specific domain. While fuzzy systems have their advantages, they may not consistently deliver the same level of performance as other machine learning methods, like deep learning, in certain tasks. The effectiveness of fuzzy logic systems can be influenced by the careful selection of membership functions and rule parameters.

To summarize, fuzzy logic serves as an invaluable asset in the realm of machine learning, enabling the handling of data that is uncertain and imprecise, as well as facilitating decision-making in scenarios where traditional binary logic falls short.

5 Applications of Fuzzy Logic

Fuzzy logic has applicability in many different industries including control systems (such as HVAC and industrial automation) where it handles imprecise inputs and optimizes system performance; automotive systems (like ABS and airbag deployment) to make real-time decisions based on uncertain conditions; consumer electronics (like washing machines and cameras) for user-friendly interface and automated adjustments; medical diagnosis systems to interpret ambiguous patient data and assist in decision- making; natural language processing to handle linguistic uncertainty in text and speech analysis; and image processing for tasks like image enhancement, segmentation, and pattern recognition, where it deals with the inherent vagueness in visual data, making it a versatile tool for addressing real-world problems characterized by imprecision and uncertainty.

5.1 Fuzzy Logic in Pattern Recognition

Fuzzy logic plays a significant role in pattern recognition by addressing the inherent uncertainty and vagueness present in many real-world recognition tasks. Here are some ways fuzzy logic is applied in pattern recognition.

Feature Extraction: Fuzzy logic can be used in feature extraction to capture and represent the uncertainty associated with the features. Instead of assigning a binary value to a feature, fuzzy sets can be used to express the degree of membership of a data point to different feature categories [5].

Fuzzy Pattern Classification: Fuzzy logic can be applied directly to the classification of patterns. In fuzzy pattern classification, class membership is expressed as a degree of belongingness rather than a binary decision. Fuzzy classifiers can assign a pattern to multiple classes with varying degrees of membership.

Fuzzy Inference in Classification: Fuzzy inference systems (FIS) can be employed for pattern classification tasks. FIS combines fuzzy rules and membership functions to determine the degree of belongingness of a pattern to different classes. This approach can handle situations where patterns exhibit characteristics of multiple classes.

Handling Noisy Data: Fuzzy logic is effective in dealing with noisy or imprecise data. It allows for the representation of uncertainty, making it more robust in noisy environments.

Fuzzy Sets for Linguistic Variables: Fuzzy sets are particularly useful for modeling linguistic variables in pattern recognition. Linguistic terms such as "very tall," "slightly cold," or "moderately fast" can be represented using fuzzy sets, allowing for a more natural and interpretable description of patterns.

Hybrid Approaches: Fuzzy logic is often combined with other pattern recognition techniques, such as neural networks or genetic algorithms, to improve recognition accuracy. These hybrid systems can leverage the strengths of both fuzzy logic and other methods.

Rule-Based Systems: Fuzzy rule-based systems can be used to express expert knowledge in pattern recognition tasks. Experts can define fuzzy rules that describe the relationships between input features and output classes, making the system interpretable and adaptable.

Image Processing: In image recognition, fuzzy logic can be used for edge detection, image segmentation, and object recognition, especially when dealing with images containing varying levels of noise or ambiguity.

Natural Language Processing: In text and document analysis, fuzzy logic can help with tasks like text categorization, sentiment analysis, and information retrieval by handling linguistic uncertainty and vagueness in textual data.

Overall, fuzzy logic provides a flexible and interpretable framework for addressing uncertainty and handling complex patterns in various pattern recognition tasks. Its ability to represent and reason with fuzzy or imprecise information makes it a valuable tool in situations where crisp, binary classification is inadequate.

5.2 Fuzzy Logic in Image Processing

Fuzzy logic plays a significant role in image processing, allowing for the handling of uncertainty and imprecision in image analysis and manipulation [9, 10]. Here are some ways in which fuzzy logic is applied in image processing:

Image Enhancement: Fuzzy logic can be used for image enhancement to improve the visual quality of images. Fuzzy-based algorithms can adaptively adjust image contrast, brightness, and sharpness based on local image characteristics and user preferences.

Image Segmentation: Different clustering algorithms are applied in image segmentation tasks. These methods allow pixels to belong to multiple clusters with different degrees of membership, making them suitable for segmenting regions with varying characteristics and partial memberships.

Edge Detection: Fuzzy logic can be used in edge detection to identify edges with varying degrees of certainty. Fuzzy edge detection methods consider gradient strengths and local characteristics to detect edges more robustly, especially in noisy images.

Image Fusion: Fuzzy-based image fusion techniques combine information from multiple images (e.g., multispectral or multimodal images) to create a single, fused image with enhanced information content. These methods take into account the uncertainty associated with each input image.

Image Denoising: Fuzzy logic can be applied to remove noise from images. Fuzzy filters adaptively smooth images by considering the local characteristics of pixel neighborhoods, which can be particularly effective in preserving edges while reducing noise.

Content-Based Image Retrieval: Fuzzy logic is used in content-based image retrieval systems to match query images with database images based on similarity degrees rather than strict binary matches. This approach allows for more flexible and robust image retrieval [11].

Image Recognition: Fuzzy logic is employed in image recognition tasks to handle the ambiguity and variability in object or pattern recognition. Fuzzy classifiers can provide degrees of confidence for object recognition, which can be valuable in situations with partial occlusion or multiple interpretations.

Medical Image Analysis: In medical imaging, fuzzy logic is used for tasks such as tumor segmentation, tissue classification, and diagnostic decision support. It allows for the modeling of uncertainty in medical images and assists in more accurate diagnoses.

Image Annotation and Labeling: Fuzzy logic can aid in annotating and labeling images by associating linguistic terms (e.g., "mostly blue," "partly cloudy") with image regions or features. This makes image descriptions more interpretable and user-friendly.

Image Compression: Fuzzy systems are employed in image compression techniques, such as fuzzy clustering-based vector quantization, to reduce data redundancy and compress images efficiently while maintaining image quality.

Quality Assessment: Fuzzy logic is used for assessing the quality of images. Fuzzy quality metrics consider multiple factors, including sharpness, contrast, and noise, to provide a more comprehensive evaluation of image quality.

In summary, A strong framework for handling ambiguity and imprecision in numerous image processing applications is offered by fuzzy logic. Its ability to handle degrees of truth and membership degrees makes it a valuable tool for enhancing image quality, segmenting images, detecting features, and performing other image analysis tasks in situations where traditional binary approaches may fall short.

6 Applications of Fuzzy Logic in Soft Computing

Fuzzy logic plays a prominent role in soft computing [12, 13], a field that encompasses various computational techniques aimed at handling uncertainty, imprecision, and approximate reasoning. The following are a few uses of fuzzy logic in the wider discipline of soft computing.

Fuzzy Control Systems: Fuzzy logic controllers (FLCs) are widely used in soft computing for managing complex, nonlinear systems where traditional control techniques may not be effective. Fuzzy control systems are employed in applications like temperature control in HVAC systems, speed control in motor drives, and autonomous navigation in robotics.

Fuzzy Approximation: Fuzzy logic is used to approximate complex mathematical functions or relationships when precise modeling is challenging. This is valuable in applications like function approximation, system identification, and curve fitting.

Fuzzy Image Processing: Fuzzy logic is applied in soft computing for image enhancement, denoising, segmentation, and object recognition. It helps manage the inherent uncertainty and imprecision in visual data.

Fuzzy Optimization: Fuzzy logic is used in optimization problems where objective functions or constraints are uncertain or vague. Fuzzy optimization techniques [14–17] are applied in engineering design, resource allocation, and decision-making under uncertainty.

Natural Language Processing (NLP): Fuzzy logic is used in soft computing approaches for natural language understanding and processing. It aids in handling linguistic vagueness and uncertainty in textual data, making it useful for sentiment analysis, information retrieval, and machine translation.

Fuzzy Control in Robotics: Fuzzy logic is applied in robotics for path planning, obstacle avoidance, and robot behavior modeling. It allows robots to adapt to dynamic and uncertain environments.

Soft Computing in Finance: Fuzzy logic, along with other soft computing techniques, is used in financial modeling and risk assessment. It helps in decision-making processes in the stock market, investment portfolio management, and credit scoring [18, 19]

Healthcare and Medical Diagnosis: Fuzzy logic-based systems are used in soft computing for medical diagnosis and patient monitoring [20]. They can handle imprecise medical data and assist healthcare professionals in decision-making.

In summary, fuzzy logic is a vital component of soft computing, addressing the challenges of uncertainty and imprecision in various applications across domains such as control systems, pattern recognition, data analysis, optimization, and decision support systems. Its ability to model and reason with imprecise information makes it a valuable tool within the soft computing paradigm.

7 Recent Advancements in Fuzzy Logic Research and Applications

Fuzzy logic, a form of multi-valued logic that deals with approximate reasoning and uncertainty, continues to be an active area of research and application. Here are four recent advancements or trends in fuzzy logic research and applications:

1. **Fuzzy Cognitive Maps**: Fuzzy Cognitive Maps (FCMs) are a type of fuzzy logic-based modeling technique that has gained significant attention in recent years. FCMs are used to represent and analyze complex systems, particularly in decision-making and risk assessment scenarios. They have been applied in various domains, such as environmental management, social sciences, and medical diagnosis.
2. **Type-2 Fuzzy Logic Systems**: Traditional fuzzy logic systems rely on type-1 fuzzy sets, which have crisp membership functions. Type-2 fuzzy logic systems, on the other hand, employ fuzzy membership functions, allowing for better handling of uncertainties and imprecision. Research in this area has focused on developing efficient algorithms and hardware implementations for type-2 fuzzy systems, enabling their application in real-time decision-making processes.
3. **Fuzzy Rule-Based Systems for Big Data and IoT**: With the exponential growth of data generated by various sources, such as the Internet of Things (IoT) devices, fuzzy rule-based systems have emerged as a powerful tool for handling and processing large volumes of uncertain and imprecise data. Researchers are exploring efficient methods for extracting meaningful patterns and insights from big data using fuzzy logic techniques.
4. **Fuzzy Logic in Autonomous Systems**: The ability of fuzzy logic to handle uncertainty and imprecision has made it a valuable tool in the development of autonomous systems, such as self-driving cars, robotics, and unmanned aerial vehicles (UAVs). Fuzzy logic controllers are being employed in these systems to handle complex decision-making processes and ensure safe and reliable operation in dynamic and uncertain environments.

These advancements and trends highlight the continued relevance and applicability of fuzzy logic that include emerging fields where fuzzy logic is proving particularly promising or ongoing efforts to refine its methodologies.

References

1. Zadeh, L.A.: Fuzzy sets. Inf. Control. **8**(3), 338–353 (1965)
2. Mamdani, E.H., Assilian, S.: An experiment in linguistic synthesis with a fuzzy logic controller. Int. J. Man Mach. Stud. **7**(1), 1–13 (1975)
3. Sugeno, M.: Industrial Applications of Fuzzy Control. Elsevier Science Publishers (1985)
4. Klir, G.J., Yuan, B.: Fuzzy Sets and Fuzzy Logic: Theory and Applications. Prentice-Hall (1995)

5. Ross, T.J.: Fuzzy Logic with Engineering Applications. John Wiley & Sons (2010)
6. Dubois, D., Prade, H.: Fuzzy Sets and Systems: Theory and Applications. Academic Press (1985)
7. Pedrycz, W., Gomide, F.: An Introduction to Fuzzy Sets: Analysis and Design. MIT Press (1998)
8. Wang, L.X., Mendel, J.M.: Generating fuzzy rules by learning from examples. IEEE Trans. Syst. Man Cybern. **22**(6), 1414–1427 (1992)
9. Dubois, D., Prade, H.: Fuzzy Sets and Systems: Theory and Applications. Academic Press (1980)
10. Yager, R.R., Filev, D.P.: Generation of fuzzy rules by mountain clustering. J. Intell. Fuzzy Syst. **2**(3), 209–219 (1994)
11. Singh, H.: Real-life applications of fuzzy logic. Adv. Fuzzy Syst. (2013)
12. Kalra, A., Tewari, G.: Nature inspired computing for optimization. In: Intelligent Systems and Applications in Computer Vision (Chapter 21). CRC Press (2023)
13. Kalra, A., Tewari, G.: Applications of Artificial Intelligence Across Industry 4.0. CRC Press. Taylor & Francis (2023)
14. Behzadian Moghadam, K., Albadvi, A.: A fuzzy logic-based decision support system for sustainable urban development planning. Sustain. Cities Soc. **45**, 330–342 (2019)
15. Kazemipoor, H., Rostamzadeh, R.: Fuzzy logic-based approach for optimizing energy consumption in smart buildings. Sustain. Cities Soc. **69**, 102883 (2021)
16. Ma, H., Zhou, L.: Fuzzy logic-based optimization of energy management in a grid-connected microgrid. IEEE Trans. Sustain. Energy **11**(2), 974–985 (2020)
17. Yang, Y., Wu, S.: Fuzzy logic-based approach for condition monitoring and fault diagnosis in manufacturing systems. J. Manuf. Syst. **64**, 1–15 (2022)
18. Mishra, A., Tiwari, S.: A hybrid soft computing model for stock market prediction using technical indicators. J. Comput. Sci. **48**, 101301 (2021)
19. Wang, Y., Lu, Z.: A comparative study of soft computing techniques for predicting stock market indices. Appl. Soft Comput. **97**, 106888 (2021)
20. Chatterjee, S., Shatabda, S.: A comprehensive survey on applications of soft computing techniques in medical diagnosis and prognosis. Appl. Soft Comput. **107**, 107431 (2021)

Healthcare Reimagined: The Impact of AI on Data Analytics

Pooja Dehankar and Susanta Das

Abstract Currently, healthcare is being revolutionized by the incorporation of artificial intelligence (AI) into healthcare data analytics, which offers significant gains in operational efficiency, personalized treatment plans, and diagnostic precision. Artificial Intelligence's Potential for Revolution in analysis of complex medical care data is examined in this chapter, with a focus on diagnostic analytics, prescriptive analytics, predictive analytics, and operational analytics. AI systems are able to generate valuable patterns from large and heterogeneous datasets through the use of deep learning (DL) and machine learning (ML) methods. This enables early disease detection, tailored medication, and effective resource management. Healthcare data analytics could greatly benefit from the use of AI, but it also confronts a number of difficulties. These include issues with data security and privacy, integrating disparate data sources, bias and fairness in AI models, and the interpretability of choices made using AI. This chapter also discusses benefits and challenges of AI in Data Analytics for Healthcare.

Keywords Artificial Intelligence · Machine learning · Deep learning · Data analytics · Predictive analytics · Data security · Telemedicine

1 Introduction

The discipline of big data analytics is gaining momentum and is starting to significantly impact the advancement of healthcare research and practices. Large amounts of heterogeneous, structured, and unstructured data generated by the present healthcare systems can now be collected, arranged, analyzed, and assimilated with the use

P. Dehankar
Ajeenkya DY Patil School of Engineering, Affiliated to Savitribai Phule Pune University, Charoli Bk. Via Lohegaon, Pune, Maharashtra 412105, India

S. Das (✉)
School of Engineering, Ajeenkya DY Patil University, Charoli Bk. Via Lohegaon, Pune, Maharashtra 412105, India

S. Mahajan et al. (eds.), *Smart Systems: Engineering and Managing Information for Future Success*, Information Systems Engineering and Management 22,
https://doi.org/10.1007/978-3-031-76152-2_2

of these instruments. The application of big data analytics to aid in the delivery of care and support illness research has lately increased; yet, adoption and further study in this field are still hindered by several basic problems with the big data paradigm [1–5].

Recently, AI-supported technologies have been increasingly deployed in healthcare facilities to improve the quality of treatment given and the effectiveness of medical resources [2–5]. Artificial intelligence (AI)-based technologies, such as machine learning, natural language processing, and intelligent robotics, present numerous opportunities for innovation in the knowledge-intensive healthcare sector [6–10].

AI's incorporation into the healthcare industry has brought in a new era for patient care and medical practice This shift results from AI systems' capacity to swiftly and accurately evaluate enormous volumes of data, offering insightful analysis that can enhance diagnosis and treatment.

Diagnostic and therapeutic procedures in healthcare have traditionally placed a great deal of reliance on the knowledge of medical specialists. The complexity of diseases, the need for more individualized therapies, and the growing amount of medical data provide the profession with several obstacles. Artificial intelligence (AI) is a transformative force because of its capacity to effectively handle these problems. The history of artificial intelligence and its use in healthcare are intertwined [11].

Medtech firms will continue to have many opportunities due to the latest technical developments in artificial intelligence, 3D printing, big data, machine learning, advanced sensors, mobile applications, and other areas. In terms of diagnosing and treating a wide range of ailments, IoT and AI are currently revolutionizing the healthcare sector. Technology improvements have led to a progressive increase in the number of linked medical devices, which has given rise to the Internet of Medical Things (IoMT), which is the Internet of Things (IoT) combined with healthcare. This allowed for the capture and transmission of medical-related data to the relevant parties whenever and wherever needed. The Internet of Medical Things (IoMT) lets healthcare organizations monitor and adjust patient behavior and health status in real time, while also helping to efficiently streamline clinical procedures, patient data, and related work flows to boost operational productivity. The delivery of P4 Medicine (Participatory, Personalized, Predictive, and Preventive) has been made possible by the Internet of Medical Things (IoMT) through the use of linked sensors and devices that provide real-time patient care, even in remote locations [12–17].

The dynamic healthcare system generates an enormous amount of diverse data. Health system management is becoming more and more dependent on intelligent and interactive analysis of health data as value-based healthcare systems take shape. In the context of improving patient outcomes and service quality while optimizing resources, this is particularly true. Innovations in big data and artificial intelligence are affecting health data analytics through innovative approaches and creative ideas [18–20].

The concept of big data goes much beyond the nature of the data; it includes aspects of data analysis, including hypothesis-generating rather than hypothesis-testing, and is commonly characterized by volume, diversity, velocity, and veracity. Frequently, big data eliminates the need for underlying probability distribution assumptions since it prioritizes the temporal stability of the link over the causal relationship. The material to be evaluated, medical big data, has a number of characteristics that set it apart from both standard clinical epidemiology and big data from other disciplines. Numerous domains in healthcare, including public health, research, disease or safety surveillance, clinical decision support, and predictive modeling, are applications of big data technology. Regression, clustering, and classification are only a few of the analytical techniques used by big data analytics that were first created in data mining. Medical big data analyses share the fundamental constraints of observational studies, including the inability to verify causality due to residual confounding and reverse causation, and are worsened by a number of technical challenges, including biased control, missing values, and the curse of dimensionality. In an attempt to get beyond these restrictions, propensity score analysis and instrumental variable analysis were developed recently, and they've made significant progress [21–27].

Quantifiable data is being used more and more in medical decision-making. Patients' whole medical history is electronically documented as soon as they interact with the healthcare system. Technology has made it easier to measure clinically significant health indicators quickly and accurately, regardless of if a patient is in a hospital ward or an operating room. This has helped to guide care and guarantee the best possible clinical outcomes. Unfortunately, as the number and complexity of biological data rise, it becomes harder to process "big data" effectively using conventional approaches. Scientists and physicians must be prepared to go beyond conventional data processing methods in order to gather knowledge from data that is therapeutically significant [26–30].

2 Machine Learning and Deep Learning Techniques Used in Healthcare Data Analytics

There is widespread use of deep learning and machines. Deep learning (DL) and machine learning (ML) are being used more frequently in the healthcare sector, particularly when combined with big data analytics. Healthcare applications for machine learning and deep learning include predictive analytics, drug development, medical image analysis, customized medicine, and electronic health record (EHR) analysis. In the area of computer science, it has grown into a very popular and highly advanced instrument. There are now more opportunities for research and development thanks to the growth of ML and DL in numerous domains.

It has the potential to transform the capacity for forecasting and making choices. One of the most important tactics in the healthcare sector has been the increasing usage of ML and DL. The biggest difficulty facing the healthcare business is dealing with a vast volume of complex, unstructured medical imaging data from gadgets, sensors, health monitoring devices, and other sources. The datasets for SCI/SCI-E/ESCI journals are taken from the WoS database and used throughout the entire investigation. Aside from this, several search strategies are applied to the extracted research documents in order to conduct the necessary scientific analysis. Bibliometrics R statistical analysis is carried out for study field, sources, documents, year, nation, and author-based analysis. Citation, co-citation, author, source, nation, institution, global collaboration, and trending word co-occurrence networks are all created using the VOS viewer program. Big data analytics paired with ML and DL have the power to transform healthcare by lowering costs, increasing the rate at which novel medicines are developed, and improving patient outcomes [30].

Predictive healthcare has been crucial in saving lives in recent years.

In the healthcare sector, intelligent systems for interpreting intricate data links and converting them into precise information for use in prediction are emerging quickly. Consequently, artificial intelligence is fast transforming the healthcare industry. Consequently, the development of processes that identify and forecast diseases, whether from clinical data or based on photographs, is increasingly dependent on systems that use machine learning and deep learning. Due to their ability to mimic human perception and even diagnose disorders that are hard for people to detect, these systems provide excellent clinical support. Predictive analytics is a critical need in the healthcare business. If a timely and correct prognosis is given, it may save patients' lives; on the other hand, an inaccurate prognosis may endanger patients' lives. This can have a substantial impact on the accuracy of illness prediction. For this reason, diseases need to be accurately predicted and estimated. Thus, it is essential to have reliable and efficient methods for predictive analysis in healthcare [30–35].

A critical task for intelligent healthcare systems utilizing machine learning (ML) is the analysis of medical data. Patients' medical histories, insurance information, and other sensitive data are typically included in the vast amounts of sensitive data that make up healthcare data. Managing and scrutinizing the data to derive significant information is a laborious procedure. Machine learning (ML) is essential for analyzing the data collected above in order to tackle this problem. Machine learning processes the medical data and predicts the outcomes by using sophisticated algorithms and a number of statistical techniques. They can be classified as supervised, unsupervised, or semi-supervised approaches to effective data analysis. The neural network under the machine learning technique plays a key role in promoting a deeper analysis of such sensitive data. Deep learning (DL) models are specifically used in the healthcare industry for pattern detection. The AI/ML paradigm combines machine learning with artificial intelligence includes DL, which imitates human learning. As a result, using this method to analyze and identify trends in clinical data is strongly advised [36–40].

3 AI and Patient Outcomes

In clinical cardiology, anatomical segmentation has been crucial. This procedure has been completely transformed by new approaches in artificial intelligence-based computer vision, which offer both automation and innovative uses. The past few years have seen a rise in study on cardiac segmentation. The most common foundation for segmentation models is a U-Net structure. There have been numerous additions made to the pre-processing or pipeline connecting aspects of analysis. The most segmented modality is cardiac MRI, partly because of relatively substantial, publicly available datasets from computer vision competitions. There has been tremendous advancement in convolutional neural networks toward the goal of cardiac anatomical segmentation. These developments lay the groundwork for automating picture analysis and for additional analysis by computer and human systems [41–45].

Significant advances in cardiovascular imaging are probably coming from data science. Timing, efficiency, and missed diagnosis are problems that affect every step of the imaging chain. The successful implementation of artificial intelligence (AI) in healthcare requires the following: reliable data; adequate computational techniques and tools; and clinical application validation for segmenting images, automated measurements, and lastly automated diagnosis. At the stages of picture capture, interpretation, and decision-making, AI may save costs and increase value. Furthermore, pathology and "big data" from the electronic health record, along with the increased precision that comes with cardiovascular imaging, should help to more accurately describe disease and tailor treatment. Diagnostic accuracy is increased by applying computer vision, an AI field that focuses on interpreting visual data to help interpret cardiac imaging modalities including CT and MRI scans. By analyzing large data sets, helping with treatment planning, and forecasting patient outcomes, cognitive computing—a rapidly developing field of artificial intelligence that mimics human cognitive functions—improves healthcare [46–50].

Emotional interactions with and via computers are the focus of affective computing, or AC. It is described as "emotion-related, emotion-arising, or purposefully influencing computing."

AC makes it possible to apply helpful and assistive technology in the medical field as well as to research and comprehend the relationship between human emotions and health. With its emphasis on identifying and addressing human emotions, affective computing is becoming increasingly important in patient-centric cardiology. Affective computing helps create more individualized treatments and better patient experiences by evaluating the feelings and actions of patients. The AI paradigm of reinforcement learning, which prioritizes trial-and-error decision-making, has the potential to enhance therapeutic treatments and treatment strategies in cardiology by promoting patient-specific, adaptable approaches [50–52]. Speech recognition software enables smooth communication between medical staff and technology, speeding up data input procedures and allowing medical records to be dictated in real time [53–55]. The combination of these several AI fields provides a multifaceted strategy to tackle the problems in cardiology, from accurate diagnosis and customized therapies to efficient workflows and patient-focused care. These developments have the potential to increase healthcare providers' productivity and improve diagnostic accuracy. Further, AI has proven to be remarkably effective at anticipating cardiac events and customizing care according to each patient's unique profile [52].

4 Sources of Healthcare Big Data

According to Rouse and Serban [44], healthcare data are inconsistent, unstructured, and come from a range of sources in a number of formats. Physiological, behavioral, molecular, environmental exposure, medical imaging, illness management, historical prescription medication histories, dietary patterns, and exercise routines are all included in the category of big data in healthcare. Big data in the healthcare sector is mostly sourced from clinical databases, electronic health record data, laboratory information system data, administrative databases (medication and insurance claims), and electronic health record data. Social media, wearable or sensor-generated biometric data, patient-reported data from standardized health surveys, medical imaging data, biomarker data, and various types of "omics" data are among the additional data sources. Miller distinguishes between two main sources of health Big Data: genomics-driven Big Data (genotyping, gene expression, sequencing data) and Big Data from payers and providers [1–3, 55–59].

Swan [24] separated streams of Big Data into two categories: (a) outdated medical data from lab reports, prescription histories, and electronic medical records that aid in better understanding infection conclusions and maximizing healthcare delivery, and (b) "Omics" data from proteomics, metabolomics, and genomics that aid in understanding the tools underlying syndromes and expediting the personalizing medical care. (c) Data obtained through wearables, sensors, and social media regarding a person's activities and way of life. Therefore, the healthcare data is provided by both internal and external data sources [42, 57].

5 Security and Privacy with Big Healthcare Data

With the recent surge of medical record digitization, the healthcare industry has seen a paradigm shift. This leads to an increase in the complexity, diversity, and timeliness of data in the healthcare industry. Big data emerges as a promising solution that might fundamentally alter the healthcare industry as medical professionals look for every way to save costs while improving the quality, delivery, and administration of treatment. This paradigm change in healthcare from reactive to proactive can ultimately contribute to economic development and a reduction in overall healthcare expenses. As the healthcare sector uses big data to its advantage, security and privacy concerns are top of mind since new risks and weaknesses keep appearing [36, 44, 45, 47, 48].

The way organizations across all industries handle, analyze, and use data has been totally revolutionized by big data. One of the most intriguing sectors to watch for changes from big data is the healthcare sector. Huge amounts of medical data have the potential to significantly improve patient outcomes, foresee epidemics, provide meaningful information, stop preventable diseases, reduce healthcare expenditures, and improve overall quality of life. Deciding which data uses are appropriate while protecting patient privacy and security may be difficult. Regardless of how valuable big data is for the advancement of medical knowledge and crucial to the success of all healthcare organizations, its use is contingent upon the resolution of security and privacy concerns [33].

Privacy has the duty to guarantee the confidentiality of each patient's unique medical information within datasets, whereas big data security includes preventing unauthorized access, use, and change of healthcare data. The aforementioned pair plays a crucial role in guaranteeing that big data is utilized efficiently for study and customized precaution, while upholding confidentiality of medical records and health data. As a result, protecting sensitive data becomes more important while utilizing big data analytics as the amount of data in the healthcare industry increases tremendously [5].

Healthcare data is gradually becoming digital, and modern healthcare uses a lot of new equipment. As a result, medical data is growing daily and has reached a significant global scale. The safety and secrecy concerns are so compelling that the health business is unable to fully utilize this data, despite the fact that it is being talked as fundamental to provide priceless insights and sinking costs. Since healthcare privacy is a key aspect governed by medical acts, it is imperative that data be protected from being misused or falling into the wrong hands. It's critical to safeguard the present big data in healthcare due to the growing risks of loss and leakage of personal information as well as the increased adoption of cloud computing [50].

6 Descriptive Analytics

Built on both past and contemporary statistics, descriptive analytics is valuable tool for learning about the past and the relationships between different factors that determine patterns by utilizing statistical parameters such as mean, range and the deviation from the mean. By utilizing methods such as online analytical processing (OLAP), descriptive analytics leverages historical data to offer insights into organizational dynamics. Descriptive analytics is frequently demonstrated by data visualization, dashboards, reports, charts, and graphs that show important business metrics including sales, orders, clients, and financial performance [55].

The descriptive analytics approach provides reports of sick or infected patients in an accessible, qualitative format for human consumption. Healthcare service providers utilize descriptive analytics to create massive amounts of data into comprehensible descriptive forms, healthcare reports of occurred measures along with resources utilized or with new clinical diagnoses for patients. Consequently, identifying patients with diabetes is not too difficult in accordance with predicted outcomes that differ from government expectations or to pinpoint areas where clinical quality measures or other parts of care need to be improved [4].

7 Predictive Analytics

Combining prediction models with intelligent alerts is a promising way to advance preventive medicine through the use of state-of-the-art analytics techniques and data-driven insights. The goal of this thorough analysis is to clarify the complex functions of intelligent alerts and prediction models in the field of preventive medicine. To begin with, risk stratification is made possible by prediction models, which analyze large datasets that include genetic, clinical, and demographic data. By identifying individuals who are more susceptible to certain diseases, healthcare practitioners can better allocate resources and prioritize preventive actions. Furthermore, these models' intelligent alerts help with early identification by informing medical professionals about individuals who exhibit risk factors or early indicators of specific illnesses. This makes it possible to take proactive preventive actions and timely interventions. Second, by utilizing past data and unique risk profiles, predictive models assist in suggesting customized preventative treatments. Personalized suggestions can be given by taking into account individual features including genetics and environmental circumstances. This gives healthcare providers the ability to provide more efficient and focused preventative treatment.

Additionally, the optimization of resource allocation is aided by the integration of intelligent warnings and predictive models. Healthcare systems can optimize the effectiveness of preventive care by concentrating attention and resources on those who stand to gain the most from it by identifying high-risk patients. These models not only enable resource allocation and individualized medication, but they also make health monitoring and surveillance easier. Intelligent alerts can be used to identify new health concerns and possible disease outbreaks by examining patterns and trends in health data [21].

8 Diagnostic Analytics

The development of an automated diagnostic system that leverages machine precision and human understanding is necessary [15]. For the diagnosis process to produce correct results while minimizing expenses, an appropriate decision support system is required. AI would help to recognize and handle it. Classifying diseases based on several factors is a complicated task for human professionals. Currently, the medical field uses a variety of AI algorithms to diagnose illnesses with accuracy. AI is a crucial component of computer science that gives computers intelligence. Learning is an essential component of every intelligent system. For physicians and surgical specialists, figuring out the potential and forecasts for health problems is a laborious effort. In certain situations, artificial neural networks (ANNs) make fast judgments on healthcare by gathering, analyzing, and identifying relevant data points [14, 58]. In the medical industry, deep learning is utilized to help specialists with the assessment of any ailment making better medical decisions as a result. The use of fuzzy logic to diagnose illnesses based on several criteria is becoming more and more common. An example of a disease that results from injury or blockage of veins in the heart is coronary sickness, which affects the amount of oxygen that reaches the heart's organs. With the help of emerging new AI approaches, fuzzy logic is constantly evolving to identify cardiac patients globally.

9 Prescriptive Analytics

The three types of analytics are descriptive, diagnostic, and predictive that concentrate on yields on automated decisions, prescriptive analytics' broad role, and knowledge acquisition for decision support systems [36]. What-if scenarios are used in this kind of analysis to explain and ascertain the cause-and-effect relationship in order to achieve optimal performance [51]. Prescriptive analytics makes a proactive decision based on the results of predictive analytics. For this reason, the likelihood of a business loss depends on the prompt detection of current business conditions and anticipated events. In addition, the lag between proactive decisions is a period of time used to plan the action.

Descriptive analytics is used at the time of the event to find the insight and provide an explanation for what transpired and why [34]. Prescriptive analytics is the most advanced type of analytics for assessing each generated choice in light of balanced goal accomplishment. Many techniques, such as heuristics, simulation, optimization, and multi-criteria decision making, as well as enablers like big data, deep learning, and cognitive computing, have an impact on the final product's quality [18]. The effectiveness and success of prescriptive analytics relies on considering hybrid data, which consists of both structured and unstructured data. Performance is also affected by the alternatives generated in the prediction phase.

Additionally, using adaptive algorithms to modify the result based on significant parameters affects the result's dependability [27].

10 Operational Analytics

Hospital systems have limited resources, but they are constantly under pressure to deliver high-quality treatment. Hospitals, however, often discover that traditional capacity management is not successful enough. One of the reasons for this is the unpredictable and volatile nature of the healthcare sector, which needs to be managed. The fact that hospitals are open-loop systems—that is, they don't use feedback to determine if their output has achieved the intended goal of their input—offers another explanation. They don't watch what comes out of the processes they manage and utilize that data to make decisions. Few effective planning or decision support systems exist in hospital systems to assist administrators in making choices [39].

As it may have multiple detrimental effects, drug shortages are always a crucial inventory management concern for healthcare organizations. To handle many concerns about the supply, management, and usage of medications, supply chain management pairs inventory control with optimization. It is difficult to determine whether there is a shortage at a hospital for a number of unpredictable reasons, such as production challenges, supply and demand issues, and raw material constraints. Hospital supply chain (HSC) organizations need to maximize income and operate their system sustainably to avoid shortage problems. This is achieved through effective inventory management. A cost-effective inventory management strategy that doesn't compromise medical services or care. Hospital supply chains that are not working well cause stock shortages due to product discontinuities, poor performance, decreased patient safety, distribution errors, and technical mishaps. Health centers can also effectively address medication shortages in their hospital supply chains by integrating machine learning (ML) applications. By quickly analyzing huge volumes of data to find patterns and disclose insights that are too complicated for the human mind, machine learning (ML) can assist health care providers in providing the right quantity at the right time, place, and cost [3].

11 Benefits of AI in Healthcare Data Analytics

Big data analysis can provide healthcare stakeholders with insights into patient health and more effective operations. The use of big data analytics offers numerous benefits. The main sources of healthcare data include EHRs (Electronic Health Records), LIMSs (Laboratory Information Management Systems), pharmacies, MDIs (Monitoring and Diagnostic Instruments), finance departments (which handle insurance claims and payments), and hospital resources. The progress of data capture devices and analytical techniques has led to the enrichment of data sources with new types of data. For example, hospitals are now collecting genetic information in EHRsBusinesses and patients alike can gain from the insightful insights found in this vast array of patient data. When applied intelligently, this knowledge can lead to incredible results. Enhanced patient care is one of the potential benefits.

Reducing medical expenditures is another benefit of big data. Big data also contributes to the prevention of pharmaceutical errors by enhancing economic and administrative performance and reducing re-admissions. Big statistics can help uncover previously unidentified linkages, underlying patterns, and insights by applying statistical analysis to large datasets. Big data and machine learning can be used to analyze human genomes and determine the most effective cancer treatment or drug combinations. Healthcare stakeholders use big data to track employee productivity and examine prior patient admittance rates as part of their business intelligence strategy. By identifying patterns and connections among this data, healthcare providers can save lives, provide better treatment, and save costs. Reduce mistakes by improving administrative and financial efficiency and lowering readmission rates.

12 Application of AI in Healthcare Data Analytics

Opportunities for worldwide infectious disease surveillance are presented by big data. They anticipate that big data will greatly improve the granularity and timeliness of currently available epidemiological information, with hybrid systems supporting more accurate infectious disease forecasts and models and supporting established monitoring systems. In the context of big data, data mining is commonly used. We are aware that data mining techniques function better when there is a substantial volume of data. The authors introduce them to machine learning and large data processing using the modern plastic surgeon as an example.

One application where algorithms show promise is data analysis. Currently available data sets span a vast quantity of information and have the capacity to reveal details about a myriad of healthcare and medical procedures. Now that computer learning, artificial intelligence, and modern computational approaches are more widely used, there are a lot more chances available. For example, artificial intelligence (AI) facilitates the conversion of data into tangible and actionable insights, which can enhance clinical decision-making, deliver high-quality patient care, adapt to emergencies in real time, and save more lives. AI also facilitates cost savings at the organizational level and leverages cash for the development of buildings and systems.

AI technologies are anticipated to help medical professionals identify probable or imminent illnesses early on, enabling quicker action. Algorithms using machine learning are showing promise in forecasting health occurrences and drawing conclusions about particular health concerns. Neural network methods have demonstrated efficacy in the detection of strokes, for instance. The algorithm examines a variety of input factors, including symptoms associated with stroke, such as paresthesia in the arm or leg, sudden confusion, altered eyesight, difficulty moving, etc. The likelihood of a stroke is calculated by analyzing this input data. Other instances of how medical data is being utilized to identify and forecast future occurrences include sepsis, surgical complications, and readmissions to hospitals [55–60, 61–63].

13 Challenges of AI in Healthcare Data Analytics

One important barrier to data analytics in underdeveloped countries is the nature of data and decision-making, because healthcare decisions are sensitive. Different from many other applications, AI and big data analytics require quick and accurate patient health information that could be life-saving. The dispersed nature and varying formats of health-care data provide challenges for artificial intelligence (AI) and analysts seeking to compile them centrally.

It takes high-quality data to produce meaningful insights. Studies have shown that data accessibility and quality are essential for data analysts to succeed in the healthcare sector. The failure of one of the most widely used big data analytics tools in epidemiology can be attributed to data from Google Trends. However, it becomes difficult for specialists to install AI devices in hospitals because of a shortage of professionals and experts in nations with limited resources. Although the government, healthcare system, and private sector recognize the potential of these technologies, few of them have the resources to purchase the AI tools required to maintain the financial viability of their enterprises. Numerous trust concerns surround how businesses may use health data. If the health care data from these nations can be used to determine their performance metrics, they are reluctant to give it. There is no acceptable accuracy rate or performance criterion for AI devices in terms of patient safety. Big data analytics and artificial intelligence are not as widely used in the healthcare industry as they could be due to private companies abusing health data for personal benefit [33–36].

14 Fairness and Bias in AI Models: An Ethical Framework Perspective

When creating AI systems, fairness is essential to ensuring that all users receive reasonable and equitable results. It entails abstaining from discrimination and making sure that all groups share equitably in the advantages and hazards of AI applications. The use of AI systems in decision-making processes that affect people or communities raises special concerns about fairness [54].

AI has many revolutionary advantages, but it can also lead to a number of issues during the system development process. Lately, participants in the AI ecosystem have begun to question the underlying ethical biases in AI-enabled agents and apps. Due to the disputed nature of ethics and the complexity of AI-enabled applications, there is conflicting research about AI ethical biases. The different theories of AI ethics and the range of ethical biases have made matters more challenging for scholars, practitioners, and policy makers. Classifying the ethical biases was done using the CRISP-DM framework [22, 54, 55].

It has been demonstrated that AI models are vulnerable to biases in the training set, particularly those that discriminate against marginalized communities. Although the industry lacks a design strategy to help brainstorm and identify potential fairness issues for AI design teams consisting of people with varied backgrounds, the demand for fairness consideration among AI solution designers is developing. In order to address this issue, the Fairness in Design (FID) framework helps AI software designers discover and look into complex fairness-related issues that could otherwise be lost. The primary fairness concepts that inform designers' thought processes and help them brainstorm metrics are what define the field of fairness in AI [60].

Ensuring equitable algorithm-guided decision making is the goal of the field of fair machine learning. Fairness has been defined in a number of formal, mathematical ways over the past ten years. Neutral models made to generate non-discriminatory predictions by restricting bias with regard to anticipated outcomes for protected identities, such gender or race—are the result of algorithmic fairness solutions [17].

The industry is rapidly pushing these solutions, and there is a great temptation to incorporate ethics into algorithm design. Because they believe the model has adequately addressed the problem of bias, practitioners in the health-care industry, where the stakes may be higher, will incorporate these models into their care. Framing fairness as a purely technological problem that can be solved by adding more data or performing correct computations is unethical, even if acknowledged technical challenges are taken into account [35].

Addressing biases, ensuring justice, and fostering accountability are critical to the ethical development of AI and data science. In order to move toward a future characterized by ethical AI development, the way forward entails promoting corporate collaboration, fighting for robust legislative frameworks, and maintaining ongoing education. The call to action pushes these ideas forward and points us in the direction of a day when technology complies with social norms and produces fair results for everybody [54].

15 Future Horizons for Healthcare-Related AI Applications

With a focus on practical implementation, interdisciplinary cooperation, and the incorporation of cutting-edge technology, the future directions for AI-driven clinical decision support are appealing. Parallel to this, AI is focusing more and more on enhancing its interpretability for robustness and transparency in order to ensure a seamless integration into healthcare workflows. Data heterogeneity, interoperability, and scalability remain persistent challenges that need to be addressed with novel methods to federated learning and decentralized architectures. Combining AI with other disruptive technologies like blockchains and the Internet of Medical Things (IoMT) offers the possibility of revolutionizing data management systems for security requirements in patient-centric care delivery. Interdisciplinary cooperation between doctors, data scientists, ethicists, experts, and lawmakers will be crucial to successfully navigating the ethical, legal, and societal repercussions of artificial intelligence in healthcare. By taking a patient-centered approach, our future research initiatives must prepare for the creation of AI-driven solutions that help patients improve health outcomes and expand equitable access to high-quality treatment. The combined power of artificial intelligence and human knowledge opens the door to a future in clinical decision support that is prepared to shatter existing barriers and usher in a new era of precision medicine and tailored healthcare delivery [40].

Big data and AI coming together to develop new drugs is evidence of humanity's never-ending pursuit of knowledge and advancement. We are getting closer to a day where diseases are better understood, therapies are more individualized, and the prospect of living longer and healthier lives is more real with each new molecule discovered, model improved, and medication created. The story's remaining chapters promise to tell of obstacles overcome, limits crossed, and the unstoppable advancement of science and technology for the benefit of humanity. The combination of artificial intelligence and big data in drug discovery is not merely a new technology trend; it is a sign of a world full of exciting and revolutionary possibilities. Taking into account the expected developments as well as the wider ramifications for the pharmaceutical sector, healthcare, and society at large, it is critical that we look ahead and wonder what the next years may bring [7].

Big Data, AI, and ML-equipped hospitals and healthcare facilities will reshape the healthcare industry. AI will offer an opportunity to cut down on the time and expense involved in drug development, as the pharmaceutical industry are always in the process of inventing new technologies. Because they can process unstructured data and integrate it with learnt data to anticipate a correct result, these artificial intelligence (AI) techniques are useful for diagnosing particular diseases. It has been demonstrated that AI-powered case-based reasoning, clinical decision support, intelligent computer-assisted instruction (ICAI), and vector regression analysis are crucial tools for monitoring the course of chronic illnesses and enhancing treatment. The method of vector regression can in handy when ascertaining the correlations among variables; Case-based reasoning facilitates problem-solving by drawing on similar prior experiences. Clinical decision support provides the healthcare team with patient-specific knowledge and information to support disease monitoring and treatment. ICAI is useful in computer-assisted instruction to patients to obtain an informative response from the patients. AI speeds up illness detection, personalizes treatments, streamlines processes, improves medical imaging, and forecasts outbreaks, all of which have a significant positive impact on patient care outcomes [10].

16 AI's Potential to Revolutionize Healthcare

An era of profound change in clinical procedures, healthcare management, and patient care is heralded by the introduction of Artificial Intelligence (AI) into the healthcare system. The potential of AI to analyze large amounts of data represents a paradigm shift in healthcare toward improving the human aspects of patient-centeredness, empathy, and understanding rather than just improving clinical decision-making and operational efficiencies. Healthcare personnel' burden is lessened and their attention is diverted toward patient-centric care thanks to administrative efficiency brought about by AI, which has the potential to revive the healthcare industry. While AI's accuracy and analytical skills in the clinical setting hold out hope for new heights in diagnostic and therapeutic precision, they also highlight the importance of human judgment and the unique subtleties of doctor-patient relationships. There are obstacles associated with this technological advancement, though. Algorithmic justice, data privacy, and ethical concerns are still at the center of the conversation, necessitating strong governance and unwavering adherence to moral standards. Due to these difficulties, navigating the complexity of AI integration requires a collaborative, interdisciplinary strategy that makes sure technical improvements are in line with the fundamental principles of healthcare [49].

One of the most revolutionary technologies of the twenty-first century is thought to be artificial intelligence (AI). One industry that has been identified as an early contender for AI revolution is healthcare. Numerous clinical and patient-facing apps that have the potential to save costs, relieve burden on medical staff, and ultimately enhance patient outcomes have already found their way into healthcare practices. But a number of worries have been expressed about the special qualities and dangers that come with AI technologies [48].

The application of artificial intelligence in the medical field is growing. The speed at which AI applications can integrate with a healthcare system will determine how much this performance impacts the medical industry, including disease diagnosis and treatment. The medical education system owes it to the next generation of physicians to equip them with the skills and knowledge required to take on roles as information integrators, interpreters, and patient advocates. The human and environmental elements impact people's confidence in AI. Long-term trust-building will be largely dependent on how well the AI performs. We are not now in a position to supervise the ethical design, development, and implementation of AI for health due to the field's rapid proliferation. Whether or not this is always the morally correct course of action, we still have a responsibility to thoroughly consider the ethical implications of deploying AI and to offer appropriate solutions. For everyone involved, striking the right balance between market forces and regulatory protections is essential because it's unclear who will ultimately be in charge of approving, overseeing, or profiting from the use of AI [29].

17 Conclusion

Today, a wide range of sources, including smart phones, wearable sensors, clinical reports, patient records, researchers, healthcare practitioners, and organizations, are producing medical data. Healthcare data may facilitate prompt decision-making in the identification of pandemics and catastrophic infections. Because of the requirement to handle and store heterogeneous, complex, and unstructured data in real-time, the growing adoption of the Internet of Things (IOT) has created challenges for analytical systems. IOT big data analytics must be integrated in order to lower healthcare costs and identify risks. The design calls for the employment of cutting-edge technology and procedures due to its incredible intricacy. Deep Learning techniques in artificial intelligence (AI) can be used to train machines to assist the medical industry in a number of ways. AI algorithms in the medical and healthcare domains could assist doctors in developing more effective patient care strategies and plans. The classic medical functions of diagnosis, prediction, and therapy are all assisted by AI.

Healthcare professionals are continuously looking for new and creative methods to improve patient care. It has embraced technological progress in order to make things better in the future. The medical field is undergoing a transformation because to big data. It is only recently that opinions concerning the delivery of healthcare have begun to shift among patients, doctors, and other healthcare providers. The use of big data as it has been discussed is just the tip of the iceberg. Because of data science's advancement and the rise of several data-driven applications, the health sector remains a leading provider of specialized services and data-driven solutions for a better life for its clients. Data scientists can use their extensive database of healthcare-related data to obtain insightful knowledge on how to increase the productivity of pharmaceutical and medical services.

Numerous areas, including mental health, public health, privacy protection, pharmacovigilance, health care management and administration, and fraud detection, are where analytics are used in the healthcare industry. Researchers have utilized data extraction to manage patients, enhance quality, save expenses, maximize resources, and carry out cloud-based computing.

Glossary

Descriptive Analytics Interpreting past data to comprehend changes that have happened in a firm is known as descriptive analytics. It facilitates the human interpretation of raw data by summarizing it. Three essential components of descriptive analytics are data visualization, data mining, and data aggregation.

Operational Analytics Business analytics that concentrates on enhancing and streamlining daily operations is known as operational analytics. It seeks to improve decision-making, expedite procedures, and boost overall efficiency through the use of real-time data and analysis.

Prescriptive Analytics Analytical research devoted to determining the optimal course of action in a particular circumstance is known as prescriptive analytics. Beyond descriptive and predictive analytics, prescriptive analytics makes targeted recommendations for decisions and actions by fusing data analysis, mathematical models, and computing algorithms.

Diagnostic Analytics Understanding the causes of historical outcomes and events is the main goal of diagnostic analytics, a type of data analysis. In-depth analysis of the causes and contributing variables of those outcomes sets it apart from descriptive analytics, which only summaries past data.

Predictive Analytics The application of statistical algorithms, machine learning methods, and historical data to forecast future events based on historical data is known as predictive analytics. Through actionable insights, it gives organizations the ability to foresee trends, comprehend possible future scenarios, and make well-informed decisions.

Healthcare Data Analytics Analytics of healthcare data is the process of examining healthcare data to extract knowledge that might enhance overall health outcomes, operational effectiveness, and patient care. To convert unprocessed data into useful information, it includes a variety of approaches and procedures.

References

1. Belle, A., Thiagarajan, R., Soroushmehr, S.R., Navidi, F., Beard, D.A., Najarian, K.: Big data analytics in healthcare. BioMed Res. Int. **2015**(1), 370194 (2015).
2. Yoon, S.N., Lee, D.: Artificial intelligence and robots in healthcare: what are the success factors for technology-based service encounters? Int. J. Healthc. Manag. (2018).
3. Holmes, J., Sacchi, L., Bellazzi, R.: Artificial intelligence in medicine. Ann. R. Coll. Surg. Engl. **86**, 334–338 (2004).
4. Safavi, K., Kalis, B.: How AI can change the future of health care. Harv. Bus. Rev. (2019).
5. Lee, D., Yoon, S.N.: Application of artificial intelligence-based technologies in the healthcare industry: opportunities and challenges. Int. J. Environ. Res. Public Health **18**(1), 271 (2021).
6. Venkatesh, D.A.N.: Reimagining the future of healthcare industry through internet of medical things (IoMT), artificial intelligence (AI), machine learning (ML), big data, mobile apps and advanced sensors. Artificial Intelligence (AI), Machine Learning (ML), Big Data, Mobile Apps and Advanced Sensors (October 28, 2019).
7. Abidi, S.S.R., Abidi, S.R.: Intelligent health data analytics: a convergence of artificial intelligence and big data. In: Healthcare management forum, vol. 32, no. 4, pp. 178–182). Sage CA: Los Angeles, CA: SAGE Publications (2019, July).
8. Lee, C.H., Yoon, H.J.: Medical big data: promise and challenges. Kidney Res. Clin. Pract. **36**(1), 3 (2017).
9. Kanevsky, J., Corban, J., Gaster, R., Kanevsky, A., Lin, S., Gilardino, M.: Big data and machine learning in plastic surgery: a new frontier in surgical innovation. Plast. Reconst. Surg. **137**(5), 890e–897e (2016).
10. Kumari, J., Kumar, E., Kumar, D.: A structured analysis to study the role of machine learning and deep learning in the healthcare sector with big data analytics. Arch. Comput. Methods Eng. **30**(6), 3673–3701 (2023).
11. Badawy, M., Ramadan, N., Hefny, H.A.: Healthcare predictive analytics using machine learning and deep learning techniques: a survey. J. Electr. Syst. Inf. Technol. **10**(1), 40 (2023).
12. Devi, K.R., Suganyadevi, S., Balasamy, K.: Healthcare data analysis using deep learning paradigm. In: Deep Learning for Cognitive Computing Systems: Technological Advancements and Applications De Gruyter, pp. 129–148 (2023).
13. Kwan, A.C., Salto, G., Cheng, S., Ouyang, D.: Artificial intelligence in computer vision: cardiac MRI and multimodality imaging segmentation. Curr. Cardiovasc. Risk Rep. **15**, 1–8 (2021).
14. Dey, D., Slomka, P.J., Leeson, P., Comaniciu, D., Shrestha, S., Sengupta, P.P., Marwick, T.H.: Artificial intelligence in cardiovascular imaging: JACC state-of-the-art review. J. Am. Coll. Cardiol. **73**(11), 1317–1335 (2019).
15. Luneski, A.N., Konstantinidis, E.V., Bamidis, P.D.: Affective medicine. Methods Inf. Med. **49**(03), 207–218 (2010).
16. Aldosari, B., Babsai, R., Alanazi, A., Aldosari, H., Alanazi, A.: The progress of speech recognition in health care: surgery as an example. ICIMTH, 414–418 (2023).
17. Dinari, F., Bahaadinbeigy, K., Bassiri, S., Mashouf, E., Bastaminejad, S., Moulaei, K.: Benefits, barriers, and facilitators of using speech recognition technology in nursing documentation and reporting: a cross-sectional study. Health Sci. Rep. **6**(6), e1330 (2023).
18. Sotirakos, S., Fouda, B., Mohamed Razif, N.A., Cribben, N., Mulhall, C., O'Byrne, A., Moran, B., Connolly, R.: Harnessing artificial intelligence in cardiac rehabilitation, a systematic review. Future Cardiol. **18**(2), 154–164 (2021).
19. Rouse, W.B., Serban, N.: Understanding and Managing the Complexity of Healthcare. MIT Press (2014).
20. Auffray, C., Balling, R., Barroso, I., Bencze, L., Benson, M., Bergeron, J., Bernal-Delgado, E., Blomberg, N., Bock, C., Conesa, A. and Del Signore, S., Zanetti, G.: Making sense of big data in health research: towards an EU action plan. Genome Med. **8**, 1–13 (2016).
21. Mohammed, E.A., Far, B.H., Naugler, C.: Applications of the MapReduce programming framework to clinical big data analysis: current landscape and future trends. BioData Min. **7**, 1–23 (2014).

22. Weng, C., Kahn, M.G.: Clinical research informatics for big data and precision medicine. Yearb. Med. Inform. **25**(01), 211–218 (2016).
23. Miller, K.: Big data analytics in biomedical research. Biomed. Comput. Rev. **2**, 14–21 (2012).
24. Swan, M.: The quantified self: fundamental disruption in big data science and biological discovery. Big Data **1**(2), 85–99 (2013).
25. Raghupathi, W., Raghupathi, V.: Big data analytics in healthcare: promise and potential. Health Inf. Sci. Syst. **2**, 1–10 (2014).
26. Ward, M.J., Marsolo, K.A., Froehle, C.M.: Applications of business analytics in healthcare. Bus. Horiz. **57**(5), 571–582 (2014).
27. Patil, H.K., Seshadri, R.: Big data security and privacy issues in healthcare. In 2014 IEEE international congress on big data, pp. 762–765. IEEE (2014, June).
28. Abouelmehdi, K., Beni-Hessane, A., Khaloufi, H.: Big healthcare data: preserving security and privacy. J. Big Data **5**(1), 1–18 (2018).
29. Al Zaabi, M., Alhashmi, S.M.: Big data security and privacy in healthcare: A systematic review and future research directions. Inf. Dev. 02666669241247781 (2024).
30. Siddique, M., Mirza, M.A., Ahmad, M., Chaudhry, J., Islam, R.: A survey of big data security solutions in healthcare. In Security and Privacy in Communication Networks: 14th International Conference, SecureComm 2018, Singapore, Singapore, August 8–10, 2018, Proceedings, Part II, pp. 391–406. Springer International Publishing (2018).
31. Vassakis, K., Petrakis, E., Kopanakis, I.: Big data analytics: applications, prospects and challenges. In: Mobile Big Data: A Roadmap from Models to Technologies, pp. 3–20 (2018).
32. Ahire, P.R., Hanchate, R.: Predictive and descriptive analysis for healthcare data. In: Handbook of Intelligent Healthcare Analytics: Knowledge Engineering with Big Data Analytics, pp. 213–231 (2022).
33. Choi, E., Bahadori, M.T., Schuetz, A., Stewart, W.F., Sun, J.: Doctor ai: predicting clinical events via recurrent neural networks. In: Machine Learning for Healthcare Conference, pp. 301–318. PMLR (2016, December).
34. Weng, C.H., Huang, T.C.K., Han, R.P.: Disease prediction with different types of neural network classifiers. Telemat. Inform. **33**(2), 277–292 (2016).
35. Menezes, B.C., Kelly, J.D., Leal, A.G., Le Roux, G.C.: Predictive, prescriptive and detective analytics for smart manufacturing in the information age. IFAC-PapersOnLine **52**(1), 568–573 (2019).
36. Sivarajah, U., Kamal, M.M., Irani, Z., Weerakkody, V.: Critical analysis of Big Data challenges and analytical methods. J. Bus. Res. **70**, 263–286 (2017).
37. Lepenioti, K., Bousdekis, A., Apostolou, D., Mentzas, G.: Prescriptive analytics: literature review and research challenges. Int. J. Inf. Manag. **50**, 57–70 (2020).
38. Delen, D.: Prescriptive Analytics: The Final Frontier for Evidence-based Management and Optimal Decision Making. FT Press (2019).
39. Khalifa, M., Zabani, I.: Utilizing health analytics in improving the performance of healthcare services: a case study on a tertiary care hospital. J. Infect. Public Health **9**(6), 757–765 (2016).
40. Munavalli, J.R., Boersma, H.J., Rao, S.V., Van Merode, G.G.: Real-time capacity management and patient flow optimization in hospitals using AI methods. In: Artificial Intelligence and Data Mining in Healthcare, pp. 55–69 (2021).
41. Abu Zwaida, T., Pham, C., Beauregard, Y.: Optimization of inventory management to prevent drug shortages in the hospital supply chain. Appl. Sci. **11**(6), 2726 (2021).
42. Rehman, A., Naz, S., Razzak, I.: Leveraging big data analytics in healthcare enhancement: trends, challenges and opportunities. Multimed. Syst. **28**(4), 1339–1371 (2022).
43. Bansal, S., Chowell, G., Simonsen, L., Vespignani, A., Viboud, C.: Big data for infectious disease surveillance and modeling. J. Infect. Dis. **214**(suppl_4), S375–S379 (2016).
44. Kanevsky, J., Corban, J., Gaster, R., Kanevsky, A., Lin, S., Gilardino, M.: Big data and machine learning in plastic surgery: a new frontier in surgical innovation. Plast. Reconst. Surg. **137**(5), 890e–897e (2016).
45. Kitsios, F., Kamariotou, M., Syngelakis, A.I., Talias, M.A.: Recent advances of artificial intelligence in healthcare: a systematic literature review. Appl. Sci. **13**(13), 7479 (2023).

46. Jiang, F., Jiang, Y., Zhi, H., Dong, Y., Li, H., Ma, S., Wang, Y., Dong, Q., Shen, H., Wang, Y.: Artificial intelligence in healthcare: past, present and future. Stroke Vasc. Neurol. **2**(4) (2017).
47. Saqib, M., Sha, Y., Wang, M.D.: Early prediction of sepsis in EMR records using traditional ML techniques and deep learning LSTM networks. In 2018 40th Annual International Conference of the IEEE Engineering in Medicine and Biology Society (EMBC), pp. 4038–4041. IEEE (2018, July).
48. Golas, S.B., Shibahara, T., Agboola, S., Otaki, H., Sato, J., Nakae, T., Hisamitsu, T., Kojima, G., Felsted, J., Kakarmath, S., Jethwani, K.: A machine learning model to predict the risk of 30-day readmissions in patients with heart failure: a retrospective analysis of electronic medical records data. BMC Med. Inform. Decis. Mak. **18**, 1–17 (2018).
49. Bertsimas, D., Dunn, J., Velmahos, G.C., Kaafarani, H.M.: Surgical risk is not linear: derivation and validation of a novel, user-friendly, and machine-learning-based predictive optimal trees in emergency surgery risk (POTTER) calculator. Ann. Surg. **268**(4), 574–583 (2018).
50. Kaur, A., Garg, R., Gupta, P.: Challenges facing AI and big data for resource-poor healthcare system. In 2021 Second International Conference on Electronics and Sustainable Communication Systems (ICESC), pp. 1426–1433. IEEE (2021, August).
51. Tatineni, S.: Ethical considerations in AI and data science: bias, fairness, and accountability. Int. J. Inf. Technol. Manag. Inform. Syst. (IJITMIS) **10**(1), 11–21 (2019).
52. Chowdhury, T., Oredo, J.: AI ethical biases: normative and information systems development conceptual framework. J. Decis. Syst. **32**(3), 617–633 (2023).
53. Zhang, J., Shu, Y., Yu, H.: Fairness in design: a framework for facilitating ethical artificial intelligence designs. Int. J. Crowd Sci. **7**(1), 32–39 (2023).
54. Corbett-Davies, S., Gaebler, J.D., Nilforoshan, H., Shroff, R., Goel, S.: The measure and mismeasure of fairness. J. Mach. Learn. Res. **24**(1), 14730–14846 (2023).
55. Marmot, M.: Social determinants of health inequalities. Lancet **365**(9464), 1099–1104 (2005).
56. Nadella, G.S., Satish, S., Meduri, K., Meduri, S.S.: A systematic literature review of advancements, challenges and future directions of AI And ML in Healthcare. Int. J. Mach. Learn. Sustain. Dev. **5**(3), 115–130 (2023).
57. Ashiwaju, B.I., Orikpete, O.F., Uzougbo, C.G.: The intersection of artificial intelligence and big data in drug discovery: a review of current trends and future implications. Matrix Sci. Pharma **7**(2), 36–42 (2023).
58. Bangash, S.H., Khan, I., Husnain, G., Irfan, M.A., Iqbal, A.: Revolutionizing healthcare with smarter AI: in-depth exploration of advancements, challenges, and future directions. VFAST Trans. Softw. Eng. **12**(1), 152–168 (2024).
59. Schönberger, D.: Artificial intelligence in healthcare: a critical analysis of the legal and ethical implications. Int. J. Law Inf. Technol. **27**(2), 171–203 (2019).
60. Zhang, J., Shu, Y., Yu, H.: Fairness in design: a framework for facilitating ethical artificial intelligence designs. Int. J. Crowd Sci. **7**(1), 32–39 (2023).

Wearable Health Technology—a Perspective

Pooja Dehankar and Susanta Das

Abstract Wearable Health Technology (WHT) has become a ground-breaking instrument in healthcare, providing previously unseen possibilities for proactive and individualized health and well-being monitoring. This article describes an overview of wearable health technology, including its uses, advantages, and drawbacks. Smartwatches, fitness trackers, and biosensors are examples of wearable technology that allows continuous surveillance of vital signs, physical activity, sleep patterns, and other health parameters. This technology gives users the ability to manage their health in real-time. Wearable technology combined with AI algorithms and sophisticated data analytics provides tailored insights and practical advice for managing chronic diseases, improving general lifestyle, and preventing illness. Wearable health technology has enormous potential, but issues with user adoption, interoperability, security, and data privacy still exist. Stakeholders including healthcare practitioners, tech developers, legislators, and consumers must work together to address these issues. WHT has full potential to transform healthcare delivery, enhance patient outcomes, and advance population health in the digital age. Healthcare sector has seen a shift due to WHT, which provides cutting-edge methods for tracking, evaluating, and improving people's health. These wearables allow for continuous real-time data collection on a variety of physiological characteristics. They range from smartwatches and fitness trackers to sophisticated biomedical sensors. The many effects of WHT on patient involvement, preventative medicine, and healthcare delivery are examined in this abstract. Wearable technology is being incorporated into medical procedures to provide proactive and individualized health management. These gadgets monitor various parameters (e.g., sleep patterns, level of physical activity, and heart rate) giving users and medical professionals insightful data on general health. Wearable data is generating more thorough understandings of people's

P. Dehankar
Ajeenkya DY Patil School of Engineering, Charoli Bk. via Lohegaon, Pune, Maharashtra 412105, India
e-mail: poojadehankar@dypic.in

S. Das (✉)
School of Engineering, Ajeenkya DY Patil University, Pune, Maharashtra 412105, India
e-mail: das29susanta@gmail.com; susanta2024das@gmail.com

S. Mahajan et al. (eds.), *Smart Systems: Engineering and Managing Information for Future Success*, Information Systems Engineering and Management 22,
https://doi.org/10.1007/978-3-031-76152-2_3

health states, enabling faster interventions and early diagnosis of possible problems. WHT encourages a move away from episodic care and toward ongoing surveillance and preventative measures. Wearables promote a sense of accountability and healthy lifestyle choices by giving people ability to actively control their personal well-being. Wearables also make it possible to monitor patients remotely, giving medical professionals more opportunities to improve patient care outside of typical clinical settings. Issues with data accuracy, privacy, and interoperability are important factors to take into account before wearable health equipment is widely used. However, continuous developments in data analytics, artificial intelligence, and sensor technologies help to overcome these obstacles and raise the general efficacy of wearable health devices. WHT encourages preventive, data-driven, and more individualized method. The integration of these devices into healthcare ecosystems, as they develop further, has possibilities to change patient outcomes, boost effectiveness of healthcare delivery, and support a proactive, wellness-focused healthcare paradigm. The chapter examines new developments in wearable health technologies as well as potential future paths. The main issues with wearable health technologies are covered in this chapter. This chapter highlights the evolution of wearable technology across time, from the earliest prototypes to the highly advanced, multipurpose gadgets that are on the market today. It examines the underlying technologies—such as wireless communication, sensors, and data analytics—that power wearable health devices and emphasizes how important they are to the continuous and unobtrusive monitoring of health. The chapter examines how wearable health technology is transforming healthcare delivery across various domains.

Keywords Wearable health technology (WHT) · Wireless communication · Sensors · Data analysis · Advancement

1 Introduction

Wearable technologies are those products that are incorporated into the human body in various ways and are typically network-connected [1]. WHT is generally understood to be a combination of sophisticated (smart) electronics and other body-adjustable accessories [2]. Wearable sensors are revolutionizing medicine and healthcare by assisting health observing beyond the confines of clinics and predicting health events [3]. By making long-term, ongoing health monitoring possible outside of clinic, wearable technologies are transforming biomedicine via digital health. Healthcare professionals are starting to use wearables for patient monitoring, having previously focused on consumer health enthusiasts. The quality and accessibility of healthcare are being revolutionized by the tremendous influx of new wearables-based analytic platforms into hospital Resource-constrained outdoor environments, at-home chronic disease management, and intensive care units. A few of the several recent developments made feasible by wearables include enhanced Alarms for the intensive care unit, remote neonatal temperature monitoring, and ongoing detection of arrhythmia outside of the clinic [3]. ML algorithms are useful for evaluating vast amounts of data collected by wearables, enabling medical professionals to see trends, forecast results, and decide how best to treat patients [4]. Two categories of learning exist: supervised and unsupervised [4]. The capacity of these algorithms to identify minute variations in health metrics—which may point to early warning indications of illness or other health problems—is one of their most noteworthy advantages. The use of ML algorithms to wearable healthcare technologies has seen a significant increase in research in recent years [5, 6].

By monitoring blood glucose meters and heart rate, for example, these algorithms have been used to forecast the start of chronic illnesses like mental health problems, diabetes, and cardiovascular disease [7, 8]. Healthcare professionals can get involved and stop the progress of these illnesses by identifying trends in the data. Additionally, MLA have shown useful in the tracking and treatment of lasting conditions comprising multiple sclerosis [9, 10]. They monitor motor symptoms and enable medical professionals to modify medicine dosages by analyzing information from wearable sensors, such as gyroscopes and accelerometers [11, 12]. Furthermore, WHT like fitness trackers and smartwatches have been used to observe mental health disorders like anxiety and depression using these algorithms.

Recent developments in wearable OS and mobile technology have made it feasible to design smartwatches that are both incredibly functional and comfortable to wear.

They are reasonably priced, with a wide variety of sensors, robust hardware, and a general-purpose operating system. It makes it the perfect tool for caring for the elderly. It's relatively easy to monitor an elderly person's basic health conditions; nevertheless, it can be challenging to identify typical wandering motion patterns, track motion activity, and use a smartwatch as an interior localization device [13].

By continuously collecting personal health information (PHI), fitness trackers may assist in enhancing physical action and lower the danger of diseases like hypertension, osteoporosis and diabetes [14]. Additionally, patients could reclaim their independence through input from doctors regarding their PHI. Pharmacists were advised to prescribe fitness trackers in a manner akin to that of medical equipment like blood pressure monitors [15]. Fitness trackers have health benefits, but their drawbacks make using them constantly a trade-off for the user [16]. Fitness trackers come in a variety of forms, including fabrics, wristbands, and clips. Their functions include communication, analysis, and "monitoring the minutiae of our everyday lives" [17].

With the capacity to evaluate biochemical indicators found in interstitial fluid, saliva, tears, and perspiration, or biofluids, dynamically and noninvasively, interest in wearing biosensors has grown significantly as a means of continuously providing real-time physiological information. More recently, innovations in non-invasive biomarker monitoring such as hormones, and metabolites have been made, along with advancements in electrochemical and optical biosensors. To enhance wearability and ease of use, a fusion of microfluidic sampling, transport, and multiplexing biosensing systems has been merged, reduced in size, and paired with flexible materials. While wearable biosensors are a promising technology, increased dependability will require a deeper comprehension of the relationships between blood analyte concentrations and noninvasive biofluids [18].

Wearable technology allows AI to analyze physiological data from patients and provide intelligent recommendations for diagnosis and treatment. Additionally, blockchain can enhance healthcare facilities through the protection of user privacy and the facilitation of global data sharing, empowering information, and ensuring dependability of information administration. Current chronic illness care models could be enhanced by fusing wearables, blockchain, and AI, shifting the focus from the hospital to the patient [19]. AI will transform the healthcare sector by combining the knowledge of physicians, druggists, and dietitians to deliver active, complete, and ongoing care for patients with prolonged illnesses. This will result in technical administration approach that encourages emergence of a in good physical shape state, slows the progress of illness, and lowers the rate of disability [20]. Greater precision in the results is made possible by the intelligent medical unit's ability to evaluate vast amount of health information with use of ML. This allows unit to identify intricate, nonlinear links between diseases and human bodies that are difficult to put into an equation [21]. Patients will receive improved care by investigating wearable technology in conjunction with AI techniques. However, timely collection and analysis of patients' physiological indicators from healthcare data while maintaining patient privacy is serious challenge in controlling chronic diseases [22]. Restrictions pertaining to data security and privacy frequently prohibit the exchange of individual health information amongst medical components. Furthermore, many prototypes and available DL systems lack testing validation and transparency. Blockchain technology may be able to help with this issue by serving as a decentralized shared ledger for information handling in a trustworthy and secure way. It also offers health data auditability and traceability [22, 23]. Through the use of smart contracts, the blockchain system controls wearable device co-chain data to safeguard patient privacy. With the use of a single consensus plugin, it also permits value transfer and cross-chain information sharing between many blockchain platforms [24]. Wearables, deep learning, and blockchain integration will raise the bar for managing chronic diseases and improve automation, data security, and privacy [25–28].

One important factor in offering a novel way to gather patient data for effective healthcare monitoring is wearable sensors. Nonetheless, a lot of medical information is made when wearable sensors are used for ongoing patient observing. Furthermore, the healthcare data provided by users are unstructured and available in vast quantities. It is difficult for the existing healthcare monitoring systems to accurately evaluate the crucial data that can be gleaned from social networking and sensor data. Additionally, massive information in healthcare cannot be processed using typical machine learning techniques for the purpose of anomaly prediction. Thus, a unique framework for monitoring healthcare is needed in order to precisely store and evaluate healthcare data and improve categorization accuracy. Using a big data analytics engine, this framework is cloud-based [29].

The absence of uniformity in data gathering and examination poses a noteworthy obstacle to use of ML algorithms to WHT. Effectiveness of ML algorithms in categorizing health issues is compromised when comparing data from a variety of wearable sensors [30, 31]. This issue was resolved and accuracy was raised by the development of cloud computing and big data analytics engines for the efficient storing and analysis of healthcare data.

Combination of wearable technology's functionality and innovation is going to propel its popularity, especially when it comes to its ability to continuously monitor factors connected to health [32]. Body-implanted or body-attached smart devices that may track physiological signals for health-related objectives are what we refer to as wearable health technologies. These wearable mobile health (mHealth) gadgets offer assistance with managing chronic diseases, making clinical decisions, and encouraging active and healthy lives [33]. If they live up to their potential, several areas of healthcare—especially those related to the rise of self-care—will probably undergo a profound upheaval. An increase in life expectancy leading to an aging population in wealthy nations [34], the growing youth population in the Global South with poor health frame [35], global urbanization and the rise of lifestyle diseases [36], and health problems arising from the start of climate change [37] are some of the demographic and environmental trends that are shifting the patients' responsibility for care. Individual self-care will be made possible by wearable health technology, which is part of a larger range of mobile health applications, in conjunction with a focus on preventive lifestyle changes [38]. Nevertheless, mHealth apps have not been able to give patients enough features that empower them, even despite the growing trend of encouraging people to take care of themselves [39]. In a similar vein, wearables' potential benefits are not yet realized due to the lack of clear evidence supporting their use as a beneficial tool for health promotion [34].

2 Literature Survey

Real-time physiological signal monitoring has fueled the development of numerous wearable and implantable technologies as the world's population continues to rise and need for healthcare access grows. Connected health (CH) denotes the emerging pattern of a technology-enabled approach to managing lifestyle and health [40]. Implicitly, it is a heterogeneous technology domain designed to offer remote and preventive therapies. In order to support health-related systems, engineering, and applications, CH is reliant on an intelligent approach, sensors, connectivity, and an internet-based digital information structure.

Wearable device demonstrated an ever-widening range of applications, including pharmacological effects, psychological therapies, heat-related sickness, COVID-19 prediction, fertility tracking, and underrepresented groups including people with rare disorders. The pandemic has sparked a trend to larger-scale, internet-based research, wherever wearable devices have improved understanding of the evolving epidemic, including predicting models and pandemic consequences. According to some research, big data gleaned from wearable device has the potential to revolutionize our knowledge of dynamics of population health and our ability to forecast health styles [41].

Data security/privacy are undoubtedly the biggest issues. It might be ideal to distinguish between people with serious medical conditions and those without serious medical conditions. Privacy concerns are generally highly serious. New reimbursement models that are suited to data-driven medicine must be established, as sustainable data-driven business models are currently unknown. It is necessary for the legal framework to be explicit and unequivocal regarding what is and is not permissible, which is not the case in many nations. Successful business concepts need a strong legal foundation [42].

Although wearables can perform various tasks (i.e., processing, collecting, storing, and transferring), there are safety/secrecy risks associated with them [43]. WHT stores data without encryption. These devices frequently lack biometric, pin, and password security, as well as the necessary authentication to access the data. There is a significant chance that private information will leak or put the user in danger of death if the devices end up in the wrong hands [44]. WHT frequently connects via Bluetooth, Wi-Fi, or NFC to smartphones or other smart devices, opening up another point of entry for the gadget. The wireless security of this device is not strong enough to withstand a brute-force attack. The information is transferred to and kept on different devices without encryption. When data is kept on a manufacturer's server or a cloud service, it is not secure. Since third-party apps lack security standards, unencrypted data may be stolen [45].

WHT has the ability to gather event-related data that a user may choose not to offer cloud service providers. As a result, consumer wearables ought to have settings that allow users to control context privacy. These settings should let users approve or reject situations in which data collecting and uploading is not appropriate, or both. Furthermore, a lot of wearables are made to be lightweight so that users don't even realize they're wearing them. The issue of context and action inference that may be made about a user from sensor data that is shared with a cloud service provider is known as context privacy. Certain wearables are equipped with sensors that enable the gathering of data about the user's background in addition to information about wearer. Head-mounted cameras and embedded microphones are two examples of devices that can record information that identifies individuals and activities unrelated to the wearer. The issue of bystander privacy pertains to safeguarding the privacy of unaffiliated persons who may be impacted by the usage of wearable technology in their environment [46].

WHT faces numerous obstacles. One of those is price. It must be reasonably priced for people to adopt it in the same way that smartphones have. Also, there's a chance that using wearable technology for medical purposes could hinder healthcare suppliers to meet the demand. Family doctors and other healthcare professionals will become even more concerned if wearable technology becomes more widespread. If wearable technology is to become as commonly used as smartphones, for example, it will need to be accepted by the general public. Furthermore, elements like design, aesthetics, and acceptance will become more crucial because of how noticeable wearable technology is [47].

The sensors' precision presents the biggest obstacle. The wearables need to have inexpensive, non-invasive sensors. Simultaneously, the measurements have to be as close to the "gold standard" as feasible (i.e., the measurements made with traditional invasive measurement methods). The sensors and wearables require the development of batteries with much longer lifespans. It might make sense to use ambient energy sources, such as the wearer's mobility and body heat, sunlight, and wind, rather than just depending on the battery. Healthcare data are extremely private, thus there needs to be tight control over how they are shared. The patient and the authorized medical personnel should be the only parties with limited access to the data, both in terms of length and intended uses. If required, the data should be encrypted before being sent over the network and stored on computers. Furthermore, all reasonable precautions should be taken by the data centers housing the servers to guard against malware infection and hacking [48].

Wearables have the potential to completely transform healthcare through encouraging health management and self-care, as well as moving treatment outside of hospitals, impacting businesses, and reinventing the industry. These sources claim that a wide range of user issues, including diminished human interaction, security, privacy, accuracy, dependability, and design, are becoming less concerning to users. The user experience can be improved, and human-centered design can progress with research on sensor materials and communication technologies. One major obstacle to the practical implementation of wearable systems is the acceptance of new technology. Although millennials are the ones adopting wearables, the older generation still finds it awkward to use and depend on them. There isn't an obvious necessity or advantage for wearables in fitness and wellness applications. Customers gripe about the design's discomfort and ugly appearance, its limited battery life, and its frequent connectivity challenges [49].

Businesses are focusing more on technology to set themselves apart from the competition as a result of advancements in electronic, mobile, and digital technologies. Rapidly evolving wearable technologies have grown in popularity and drawn attention from companies looking to expand into new industries and build cutting-edge goods. Wearable technologies, also known as technologies integrated into clothing and accessories worn by humans, have an impact on many different industries, but the health industry is one that is particularly influenced. It allows for the proper follow-up of an individual's medical status by transferring private data to smart devices using smart detectors. With their application targeted at boosting community wealth, wearable health technology items are becoming more and more of a necessity than a personal preference. Thus, the study centers on wearable technology goods that grab consumers' attention and firms looking to expand into new areas in order to boost sales. Examining and providing information about WHT goods and their market, that has a positive impact on community health in the wearable technology product market, is the aim of this study. Information on wearable technology and wearable health technology development is the study's main focus. As a result, it is evident that the market for WHT is expanding quickly. Businesses will have more market potential if there is a rise in request for goods that could improve people's class of life, provide quick health monitoring, and make their lives easier. As a result, the market for WHT is thought to be particularly appealing to manufacturers and customers alike [50].

The ultimate purpose of these gadgets and applications is to give patients real-time feedback so they can see data showing how harmful behaviors affect their bodies, which will further encourage them to take more active approach to managing their wellbeing. Patient will be more equipped to make significant behavioral adjustments the more informed they are about their own health. Access to, interpretation of, and conversion of customized health data into practical teaching points are requirements for healthcare providers. In the past, doctors frequently taught patients about fitness and health in an authoritarian manner that did nothing to promote patient engagement. Through the implementation of a patient-centered strategy that leverages data obtained from wearable devices, healthcare providers may collaborate with patients to impart measurable skills and habits in a concrete and real-world manner, all while monitoring progress over time. Compliance and satisfaction rates rise when patients participate in a cooperative and patient-centered conversation about their health and well-being, resulting in a more fruitful doctor-patient relationship. Recalling that these devices are still in the early phases of development, and in order to offer the finest medical advice, the validity and accuracy of the data still needs to be improved [51].

A major worry for most consumers is data security and privacy in terms of utilizing WHT to gather health data. Today, there is a greater chance of data stealing or improper use of health data due to users' constant connectivity. Moreover, many users of wearable technology are ignorant of who the data's owner is, what uses it, and who can access it. Data ownership is an often-misunderstood problem in the field of WHT. In an effort to make data privacy and ownership clearer, many nations are revisiting their privacy governance [52].

Wearable computer applications demand careful consideration of a variety of design considerations. To address hardware and software issues, network solutions, user interfaces, interaction strategies, and potential trade-offs must all be properly planned and implemented at the same time in order to satisfy system and application requirements. The process of designing a novel application includes decisions about the form factor design of the device, its physical attributes, its target users (direct and indirect beneficiaries), the usability of the interfaces considering the users' cognitive load and mental load, social impacts, and ergonomics. Stakeholders must overcome several obstacles in order to identify suitable design options for innovative applications [53].

Understanding advanced solutions in field to use again and expand becomes difficult because present solutions are fragmented and reported in disparate places. Furthermore, current solutions are typically specialized, meaning they target particular qualities, individual applications, and tiny user samples; they are not always generalizable or adaptable to different use cases. In addition, trade-offs must be made between criteria from technologists (such as hardware assembly, battery life, and device dimension), users (such as ergonomics, comfort, and aesthetics), and healthcare professionals (e.g. data accuracy, privacy, and measurement frequency). When one considers that the creation of innovative wearable solutions is on a shifting terrain, further difficulties become apparent. E-textiles, wearable technology, and smart apparel are developing swiftly. As a result, stakeholders cannot be effectively guided or supported throughout the design process by means of standardized techniques or well verified procedures. Finally, because these technologies are typically expensive to acquire and operate, care must be made to avoid making inequality worse [54].

There is great promise for healthcare in common and behavior modification in specific from integration of wearable technology with mobile applications. Mobile applications let users analyze and interpret the data acquired, increase awareness, and motivate behavior changes, while wearables gather facts and numerical details of their day-to-day existence.

Numerous uses have demonstrated efficacy in promoting healthier eating habits and integrating physical activity into users' everyday routines. Though promising, the creation of wearable healthcare solutions still faces several obstacles. In addition to the need for the devices to be user-friendly and comfortable, their underlying methods also need to be properly thought out in order to guarantee users' acceptance, adoption, and continued engagement [55].

Wearable technology has the potential to enhance self-management decision-making by regularly offering input on physiological processes. Pregnancy outcomes are eventually improved by making educated choices/decisions to assist self-control of this illness when expecting. Fitbit activity tracker, interstitial glucose monitor, and many more are among the wearable devices being considered. Apart from these gadgets, the participants would have to keep a meal journal. Participants will be requested to clearly document the kind and amount of meals consumed using cameras on their mobile phones in order to aid in the assessment of portion sizes. The gathered data will be subjected to statistical and computational analysis, following which ML techniques will be used to provide feedback regarding necessity of insulin or carbohydrates to maintain euglycemia [56].

A very common visual damage that impairs people's skill to perceive color is color blindness. Chroma, a wearable augmented reality system built on Google Glass, is there for solving this problem by letting users view a real-time, filtered picture of the current scene. Based on type of color blindness, Chroma automatically modifies the scene-view and includes specific algorithms for color saliency. Chroma can help colorblind people recognize colors better in a range of real-world situations, according to preliminary lab testing. When Chroma is implemented on a wearable augmented reality gadget, it becomes a useful tool that can enhance daily tasks and give colorblind individuals access to many color dimensions [57].

Simple body-worn sensors are being used to monitor and classify human activities, and this is becoming a significant field of machine learning research. Numerous mobile and ubiquitous computing applications, including medication assessment, medical behavioral monitoring, smart information delivery, recording systems for industrial assembly and maintenance, and assistive technologies for aged and intellectually challenged, are driving forces behind activity monitoring itself. Motion sensors positioned on the body are solely affected by user activity. Reliability of activity recognition with such sensors is less dependent on useful feature extraction. Therefore, relatively precise trajectories could be obtained for activity recognition once a visual system has been able to monitor them. Accelerometers worn on the arms respond to variations in arm speed as well as earth gravity [58].

Individuals wear gadgets on different portions of their bodies based on personal taste. Alterations depending on these are typically rough; e.g. gadget may be found inside a purse, an arm or hip holster, a jacket pocket, or pocket of pant. While on-body placement changes are less common than within-body displacements or orientation changes, they can still happen because multiple devices are linked to specific body placements (e.g., glasses are linked to the eyes); for example, a user may carry her glasses in her bag or put them in her shirt pocket. The body part on which the device is positioned can affect the signals from motion sensors in two ways: first, certain activities are linked to particular body parts and sensors, like activities involving small arm motions like hand washing; and second, the motion sensor signals vary significantly from body location to body location, even for activities that aren't completely body-part specific [59].

People who suffer from Autism Spectrum Disorder (ASD) frequently exhibit stereotyped behaviors. Some people exhibit these habits frequently, which can be disruptive and even self-destructive. The authors suggest a system that can connect instances of stereotypies to contextual data that is implicitly gathered about a person's physiological condition and their external environment. Information needs of the stakeholders were ascertained and a greater understanding of the nuances involved in satisfying those requests was provided via a system prototype created through participatory design [60].

WHT presents obstacles for the healthcare system. Primary issue faced by healthcare system is to facilitate utilization of these tools by altering care model and exchanging data. Healthcare professionals and patients must work together to integrate these technologies, not only throughout the adoption phase but also during the development/implementation of best practices and care pathways. Quality of care will increase with further study and growth of methods that will contribute to the body of knowledge and add new information while figuring out how best to apply it. By educating the patient and giving him access to resources, data, and information, it will involve the doctor and acknowledge him as a co-participant in the healthcare process. As a result, patient data from scattered systems may be collected and analyzed, and information from the patient and other sources can be shared. Regulating data use, secrecy, and safety must also be changed in order to achieve this. The difficulty with wearable technologies is to create support systems that will increase acceptance while taking into account privacy and security concerns, standardization limits, and current care paradigms. The solutions should be tailored to the system while considering the workflows and workloads of healthcare professionals. The healthcare industry must create new business models and more affordable solutions to enable large-scale adoption of wearable technology by healthcare organizations [61].

WHT makes it possible to continuously monitor physiological and biochemical markers during daily living, and bodily actions and behaviors of people. Vital indicators include blood oxygen saturation, body temperature, blood pressure, heart rate, posture, and physical activity, are primary monitored statistics using instruments such as electrocardiograms (ECG), ballistocardiograms (BCG), and others. It is possible that wearable cameras or video recorders could offer extra clinical data. It is possible to wear wearable electronics on watches, gloves, clothes, shoes, eyeglasses, and earrings. Additionally, wearable technology may develop into skin-attachable technology. It is possible to integrate sensors into furniture including chairs, car seats, and mattresses. Information is usually gathered via a smartphone and sent to a distant server for processing and storage. There are two main categories of wearable technology that are employed in the research of gait patterns. Healthcare practitioners can now analyze walking patterns with the use of gadgets like gyroscopes, multi-angle video recorders, and accelerometers. Additional health-related devices have been created, such as mobile phone apps and add-ons and on-wrist activity monitors like Fitbit. Gait assessment activities in numerous sceneries are regularly conducted in concurrence with wearable devices and data processing procedures [62].

Science is faced with the challenge of creating sustainable health care systems, and the best method to deal with this problem is by using information and communication technology. Thanks to recent developments in sensor networks, it is now possible to provide independent, anticipatory, and proactive health care services, which have been shown to be very helpful. Health-care solutions have significantly improved since the advent of Health records and web-based platforms. With the rise of numerous smartphone apps for physiological state monitoring, self-monitoring and diagnoses are now feasible [63]. These systems frequently have problems with scalability, security, and privacy, even though they are an essential step toward enabling personalized health care. Rather than providing ongoing health monitoring, these solutions are limited to addressing short-term physiological problems [64].

People these days are highly interested in wearable technology since it's the latest trend in tracking daily activities. The most advanced activity life monitors available today are pedometers' cousins. They can do much more than just figure out how long you walk; they are more intelligent, accurate, and sophisticated [65].

Wearable technology is a new kind of gadgetry that consists of a tiny piece of hardware with an application for measuring and recording fitness parameters like distance run or walked, heart rate, and sleep tracking, among other things. Nowadays, phrase is mostly used to describe specialized electronic surveillance tools that are synchronized—often wirelessly—with a computer or smartphone to follow data collected over a long time. Additionally, certain smartphones have the capacity to track on their own [66]. Wearable gadgets are small, cutting-edge computers that people can wear on different areas of their bodies, including smart watches, eyewear, bracelets, or wristbands [67, 68].

Wearables that have been created thus far can be used on any part of the human body and are divided into three groups: devices for the head, limbs, and torso [26]. The primary categories of head wearable are helmets, headbands, spectacles, hearing aids, ear buds, earrings, and patches [49]. Utilizing Google Glass for surgical navigation, medical education, and telemedicine. It can also take pictures, make video calls, and track GPS positions. Technologies for mixed reality and AR/VR can be applied to it [69]. Majority of limb wearable are those that are worn on the arms, legs, and feet [70, 71]. Most of the gadgets worn on the upper limbs are connected wristbands, smart watches, and other accessories that track physical data such as body temperature, heart rate, UV exposure level, and regular activities [72]. Most of the gadgets worn on the lower limbs are mainly used in the field of rehabilitation and come in the form of movement-related shoes and socks [73]. The majority of torso wearables are undergarments, belts, and suits [74, 75]. Rapid advancements in material science and sensor technologies have made it easier to produce electronic goods integrated in textiles or other materials that can be useful to a variety of biological uses in recent years [76].

Numerous wearables for consumers make the claim to track mental strain, and numerous researchs have used physiological signals to distinguish between experimental and control stress situations in study subjects [77–83]. Wearable technology has not yet advanced to the point where it can accurately identify whether a person is experiencing a stress response, and little evidence of external validity has been presented. Still, scientists are improving protocols for connecting digital phenotyping in naturalistic settings with stress detection in lab-based studies [84]. We might be able to recognize stress reactions quite accurately with more investigation.

Wearable technology is a vital component of customized medical care. Because of their rising levels of accuracy and capability, their use is gradually growing. Customers use them for a variety of purposes, such as recording health information, measuring physical activity, or simply rivalry amongst friends according to activity data. Different user groups have different goals in mind, such as tracking their sleep, fertility, weight loss, or gait. Smart wearables' capabilities have been steadily improving in recent years, allowing for massive data collecting. Health experts can use this data to assist with consumer education, medical diagnosis, and treatment can use it to help motivate themselves to accept and monitor healthy everyday habits [85].

The public's attention is focused on health issues due to the unusual epidemic and rise of an old population. In order to effectively identify relevant elements affecting adoption of WHT, additional settings are needed to be adequately included in studies on technology adoption, given the differences between healthcare and non-healthcare information technology. In order to create a health-aware acceptance model, we therefore carefully rearranged elements from the privacy, health, and sociodemographic settings. After that, structural equation modeling was used to conduct an empirical analysis. According to the findings, health concerns had little bearing on the intention to utilize wearable healthcare technology, but privacy issues did. Furthermore, social influence and enabling environments were moderated by age. These results provide important new information about how to boost the rate at which healthcare technology is adopted by protecting personal data, encouraging social interaction among users and providing an easy-to-use interface for senior citizens [85].

Information technology has advanced to the point that developers have enhanced the devices' application in a number of situations [86–88], such as education [89], healthcare [90], and sports [91]. One such development is the creation of wearable medical technology, which helps to improve health by encouraging sports and physical activity [92]. In general, wearable technology for healthcare refers to electronic devices that are integrated into wearable accessories [93, 94]. Examples of such devices are Fitbit, Mi Band, iWatch, and Google Glass wristbands, that are used in medical/healthcare sectors. As a result, it is thought to be among the most significant technologies of the future for raising healthcare quality and lowering costs [95].

WHT solutions are helpful in tracing information about users' health in real-time by combining technological and health aspects [96]. Wearing wristbands, for instance, allows users to track their heart rate, sleeping patterns, and distance traveled in real-time. This information may be used to analyze and manage real-time data and help users start taking action to improve their health [95].

There aren't much empirical research examining the elements that result in acceptance of WHT, despite these encouraging advancements. The objective was to acquire information in this field and offer significant management consequences for doctors and wearable device creators, with aim of filling the previously indicated gaps [97].

Future research ought to address a number of difficulties such as elements that could significantly affect user experience but are tied with the practical aspects of WHT. One of these technical aspects can help provide useful design and engineering suggestions for future study, as variations in the devices' weight, raw material composition, and size of the display screen may have an impact on user satisfaction and how devices are used. More attention should be paid to structural and functional elements that may draw senior users. Second difficulty is it's possible that the results cannot be applied broadly to explain wearable gadget adoption outside of the healthcare industry. Users of wearable medical technology are probably more driven and aware of their own health, which means they have a specific reason for buying the equipment. As such, the acceptance pattern might not apply to non-power users or wearable device usage for hedonistic or entertainment purposes. Future research endeavors will be capable of expanding upon findings and putting out a more comprehensive model that is technically and psychologically sound by tackling these limitations [98].

Wearable technology's usability may affect consumers' knowledge of data leaks and location tracking. Consumers who were aware of data leaks and location tracking expressed less happiness with wearable technology. Users who are less satisfied with the usability of a product appear to be more privacy conscious. The relationship between usability and privacy perceptions is shown by this study. Extent to which consumers understand data gathering activities may depend on the usability. It might also indicate that consumers are already aware that specific kinds of data are being collected, and that the device's poor usability is the only reason for their decreased level of happiness. On the other hand, people who were informed about the methods used to acquire health data rated usability higher and more favorably [99]. Adoption rates might be impacted by cultural attitudes, beliefs, and perceptions around wearable technology and medical procedures. It's critical to address stigma and advance acceptance. People from different cultural origins have quite different intentions when it comes to accepting these devices [100]. When it comes to integrating wearable health technology seamlessly with current healthcare systems and realizing its full potential, interoperability issues are a major obstacle. It is true that problems with interoperability can make it more difficult to incorporate wearable technology into the current healthcare systems. Proprietary data formats, unsuitable communication protocols, security concerns, and the difficulty of integrating disparate devices into an established healthcare IT infrastructure are some of the elements that give birth to these challenges. It will involve cooperation between stakeholders, standardization of communication protocols and data formats, installation of safe data exchange systems, and creation of interoperability frameworks and guidelines to overcome these interoperability obstacles. The promise of wearable health technology to advance healthcare delivery, improve patient care, and enhance clinical processes, it is imperative that these concerns be addressed. Users and medical professionals are unable to obtain a combined perspective of fitness and health data due to interoperability issues and data silos [101]. In comparison to common technologies (e.g., smartphones), uptake of these gadgets is rather sluggish. As a result, producers and designers are becoming more curious about what influences consumers' decision to use new technology. In order to impress and win over customers, they will be able to enhance the features and appeal of these products with their assistance [102].

3 Conclusion

The global effort to improve healthcare access and quality will unavoidably need wearing wearable technology to monitor people continuously and vigilantly for advantages. Research on wearable can drive the clinical revolution of healthcare delivery in the future. Wearable technology has great promise for bending the healthcare cost curve. In addition to facilitating quick diagnosis and event detection, this varied and widely distributed data source from both individuals and large populations will encourage the creation of new artificial intelligence techniques for the prediction of future healthcare conditions for everyone. A new group of computer scientists committed to developing the core ideas for wearable device computing for healthcare will be drawn to this shift in healthcare.

To fully utilize wearable health technology in healthcare and overcome the many difficulties that lie ahead, coordination among stakeholders is vital. Through cross-disciplinary collaboration, user-centered design, data integration and interoperability, clinical utility validation, regulatory compliance, ethical considerations, education, and training, and more, stakeholders can jointly shape a future in which wearable technology empowers people, enhances patient outcomes, and revolutionizes healthcare delivery.

Wearable technology needs further research, but it has a lot of potential to help anticipate clinical outcomes. To fully take advantage of the opportunities presented by this rapidly evolving technology, well-designed clinical trials that integrate wearable technology data into clinical outcome prediction models are necessary. This article attests to the growing maturity and establishment of wearable technology applications. The article should advance the transmission of useful recommendations in a sustainable manner, enforce privacy by design, provide seamless monitoring, enable ubiquitous sensing that uses less energy and makes 5G communications possible with low latency. Future initiatives including all parties involved generating scientific evidence on enhanced routes for care and their financial viability is required to supplement technological advancements. Future digital technology design and development in this field will depend on ongoing examination of best practices, problems, and viable remedies to address current issues. This article represents developments and consequences in order to incorporate patient data into Electronic Health Record, even though it does not offer an exhaustive list of all WHT efforts. Our work lays the groundwork for organizations in the health care sector to get materials on WHT adoption and procedures. WHT with AI capabilities has enormous potential to improve healthcare through instantaneous input. To enable prompt interventions and possibly life-saving measures, this can enable people to embrace improved lives and become alert of initial health risks, e.g., sleep difficulties or abnormal heart rhythms. Furthermore, employing these wearables to aggregate real-time data from a sizable population provides important means for people health research and advancements in medicine. A deeper comprehension of health developments, illness pattern, and efficacy of therapies can be attained with the use of such data-driven insights. Smartphones and wearable technology are two examples of widely used, internationally adopted technology that is only predicted to grow. This creates new difficulties for clinicians when evaluating patient data, especially when it comes to analyzing rhythm and rate in patients who are asymptomatic. It is recommended that healthcare professionals are cognizant of both the advantages and possible uses of this technology in addition to the existing limitations and concerns. To elucidate the therapeutic significance of wearables and the medico-legal and ethical ramifications of arrhythmia recognition, more research and recommendations are needed. Healthcare wearable technology gadgets are anticipated to offer fresh approaches to addressing health-related concerns, given the encouraging advancements in cutting-edge technology items and applications.

For instance, combining wearable medical technology with mobile apps which track physical activities can motivate people to engage in more physical activities, which help them maintain their health and wellness and lessen the negative effects of their sedentary lifestyle. In order to look into the antecedents of healthcare wearable technology adoption intention, we experimentally evaluated a research model. According to various research, apparent usefulness influences acceptance intention—was significantly impacted by health beliefs and the correctness of health data.

A turning point in development of personal health management has been reached with the introduction of WHT. The present chapter has examined numerous uses, advantages, difficulties, and potential future developments of wearable medical technology.

Since its inception, wearable health technology has evolved from simple fitness trackers to intricate instruments that can track vital signs continually, identify abnormalities, and give users and healthcare professionals useful information. Wearables improve patient participation, promote health outcomes, and streamline healthcare delivery, from eldercare and chronic illness management to preventative care and rehabilitation.

By empowering people, optimizing clinical workflows, boosting health outcomes, and spurring changes in healthcare delivery models, WHT has the ability to completely change the healthcare industry. In order to fully realize wearables' transformative potential, interoperability issues must be resolved, data security and privacy must be guaranteed, stakeholder engagement must be encouraged, and technological improvements must be taken advantage of in order to provide patient-centered, value-based care.

References

1. Thierer, A.D.: The internet of things and wearable technology: addressing privacy and security concerns without derailing innovation. Adam Thierer, Internet Things Wearable Technol.: Addressing Priv. Secur. Concerns Derailing Innov. **21** (2015)
2. Wright, R., Keith, L.: Wearable technology: If the tech fits, wear it. J. Electron. Resour. Med. Libr. **11**(4), 204–216 (2014)
3. Dunn, J., Runge, R., Snyder, M.: Wearables and the medical revolution. Pers. Med. **15**(5), 429–448 (2018)
4. Choi, R.Y., Coyner, A.S., Kalpathy-Cramer, J., Chiang, M.F., Campbell, J.P.: Introduction to machine learning, neural networks, and deep learning. Transl. Vis. Sci. Technol. **9**(2), 14–14 (2020)
5. Yu, K.H., Beam, A.L., Kohane, I.S.: Artificial intelligence in healthcare. Nat. Biomed. Eng. **2**(10), 719–731 (2018)
6. Harrer, S., Shah, P., Antony, B., Hu, J.: Artificial intelligence for clinical trial design. Trends Pharmacol. Sci. **40**(8), 577–591 (2019)
7. Baig, M.M., GholamHosseini, H., Moqeem, A.A., Mirza, F., Lindén, M.: A systematic review of wearable patient monitoring systems–current challenges and opportunities for clinical adoption. J. Med. Syst. **41**, 1–9 (2017)
8. Mody, V., Mody, V.: Mental health monitoring system using artificial intelligence: a review. In: 2019 IEEE 5th International Conference for Convergence in Technology (I2CT), pp. 1–6. IEEE (2019)
9. Belić, M., Bobić, V., Badža, M., Šolaja, N., Đurić-Jovičić, M., Kostić, V.S.: Artificial intelligence for assisting diagnostics and assessment of Parkinson's disease—a review. Clin. Neurol. Neurosurg. **184**, 105442 (2019)
10. Raghavendra, U., Acharya, U.R., Adeli, H.: Artificial intelligence techniques for automated diagnosis of neurological disorders. Eur. Neurol. **82**(1–3), 41–64 (2020)
11. Mukhopadhyay, A., Sumner, J., Ling, L.H., Quek, R.H.C., Tan, A.T.H., Teng, G.G., Seetharaman, S.K., Gollamudi, S.P.K., Ho, D., Motani, M.: Personalised dosing using the CURATE. AI algorithm: protocol for a feasibility study in patients with hypertension and type II diabetes mellitus. Int. J. Environ. Res. Public Health **19**(15):8979 (2022)
12. Zhu, X., Huang, W., Lu, H., Wang, Z., Ni, X., Hu, J., Deng, S., Tan, Y., Li, L., Zhang, M., Qiu, C.: A machine learning approach to personalized dose adjustment of lamotrigine using noninvasive clinical parameters. Sci. Rep. **11**(1), 5568 (2021)
13. Chun, J., Dey, A., Lee, K., Kim, S.: A qualitative study of smartwatch usage and its usability. Hum. Factors Ergon. Manuf. Serv. Ind. **28**(4), 186–199 (2018)
14. Shih, P.C., Han, K., Poole, E.S., Rosson, M.B., Carroll, J.M.: Use and adoption challenges of wearable activity trackers. IConference 2015 Proc. (2015)
15. Mercer, K., Giangregorio, L., Schneider, E., Chilana, P., Li, M., Grindrod, K.: Acceptance of commercially available wearable activity trackers among adults aged over 50 and with chronic illness: a mixed-methods evaluation. JMIR Mhealth Uhealth **4**(1), e4225 (2016)

16. Wieneke, A., Lehrer, C., Zeder, R., Jung, R.: Privacy-related decision-making in the context of wearable use (2016)
17. Newell, S., Marabelli, M.: Strategic opportunities (and challenges) of algorithmic decision-making: a call for action on the long-term societal effects of 'datification.' J. Strateg. Inf. Syst. **24**(1), 3–14 (2015)
18. Kim, J., Campbell, A.S., de Ávila, B.E.F., Wang, J.: Wearable biosensors for healthcare monitoring. Nat. Biotechnol. **37**(4), 389–406 (2019)
19. Xie, Y., Lu, L., Gao, F., He, S.J., Zhao, H.J., Fang, Y., Yang, J.M., An, Y., Ye, Z.W., Dong, Z.: Integration of artificial intelligence, blockchain, and wearable technology for chronic disease management: a new paradigm in smart healthcare. Curr. Med. Sci. **41**(6), 1123–1133 (2021)
20. Silva, P., Jacobs, D., Kriak, J., Abu-Baker, A., Udeani, G., Neal, G., Ramos, K.: Implementation of pharmacogenomics and artificial intelligence tools for chronic disease management in primary care setting. J. Pers. Med. **11**(6), 443 (2021)
21. Lu, L., Zhang, J., Xie, Y., Gao, F., Xu, S., Wu, X., Ye, Z.: Wearable health devices in health care: narrative systematic review. JMIR Mhealth Uhealth **8**(11), e18907 (2020)
22. Jiang, W., Majumder, S., Kumar, S., Subramaniam, S., Li, X., Khedri, R., Mondal, T., Abolghasemian, M., Satia, I., Deen, M.J.: A wearable tele-health system towards monitoring COVID-19 and chronic diseases. IEEE Rev. Biomed. Eng. **15**, 61–84 (2021)
23. Dwivedi, A.D., Srivastava, G., Dhar, S., Singh, R.: A decentralized privacy-preserving healthcare blockchain for IoT. Sensors **19**(2), 326 (2019)
24. Kalid, N., Zaidan, A.A., Zaidan, B.B., Salman, O.H., Hashim, M., Muzammil, H.J.J.O.M.S.: Based real time remote health monitoring systems: A review on patients prioritization and related "big data" using body sensors information and communication technology. J. Med. Syst. **42**, 1–30 (2018)
25. Qadri, Y.A., Nauman, A., Zikria, Y.B., Vasilakos, A.V., Kim, S.W.: The future of healthcare internet of things: a survey of emerging technologies. IEEE Commun. Surv. Tutor. **22**(2), 1121–1167 (2020)
26. Koydemir, H.C., Ozcan, A.: Wearable and implantable sensors for biomedical applications. Annu. Rev. Anal. Chem. **11**, 127–146 (2018)
27. Xie, Y., Zhang, J., Wang, H., Liu, P., Liu, S., Huo, T., Duan, Y.Y., Dong, Z., Lu, L., Ye, Z.: Applications of blockchain in the medical field: narrative review. J. Med. Internet Res. **23**(10), e28613 (2021)
28. Zheng, X., Sun, S., Mukkamala, R.R., Vatrapu, R., Ordieres-Meré, J.: Accelerating health data sharing: a solution based on the internet of things and distributed ledger technologies. J. Med. Internet Res. **21**(6), e13583 (2019)
29. Ali, F., El-Sappagh, S., Islam, S.R., Ali, A., Attique, M., Imran, M., Kwak, K.S.: An intelligent healthcare monitoring framework using wearable sensors and social networking data. Futur. Gener. Comput. Syst. **114**, 23–43 (2021)
30. Bayoumy, K., Gaber, M., Elshafeey, A., Mhaimeed, O., Dineen, E.H., Marvel, F.A., Martin, S.S., Muse, E.D., Turakhia, M.P., Tarakji, K.G., Elshazly, M.B.: Smart wearable devices in cardiovascular care: where we are and how to move forward. Nat. Rev. Cardiol. **18**(8), 581–599 (2021)
31. Nweke, H.F., Teh, Y.W., Al-Garadi, M.A., Alo, U.R.: Deep learning algorithms for human activity recognition using mobile and wearable sensor networks: State of the art and research challenges. Expert Syst. Appl. **105**, 233–261 (2018)
32. Casselman, J., Onopa, N., Khansa, L.: Wearable healthcare: lessons from the past and a peek into the future. Telematics Inform. **34**(7), 1011–1023 (2017)
33. Hughes, A., Shandhi, M.M.H., Master, H., Dunn, J., Brittain, E.: Wearable devices in cardiovascular medicine. Circ. Res. **132**(5), 652–670 (2023)
34. Li, C., Lin, S.H., Chib, A.: The state of wearable health technologies: a transdisciplinary literature review. Mob. Media Commun. **9**(2), 353–376 (2021)
35. Fatusi, A.O., Hindin, M.J.: Adolescents and youth in developing countries: health and development issues in context. J. Adolesc. **33**(4), 499–508 (2010)

36. Goryakin, Y., Rocco, L., Suhrcke, M.: The contribution of urbanization to non-communicable diseases: evidence from 173 countries from 1980 to 2008. Econ. Hum. Biol. **26**, 151–163 (2017)
37. Barnes, P.W., Williamson, C.E., Lucas, R.M., Robinson, S.A., Madronich, S., Paul, N.D., Bornman, J.F., Bais, A.F., Sulzberger, B., Wilson, S.R., Andrady, A.L.: Ozone depletion, ultraviolet radiation, climate change and prospects for a sustainable future. Nat. Sustain. **2**(7), 569–579 (2019)
38. Chib, A., Lin, S.H.: Theoretical advancements in mHealth: a systematic review of mobile apps. J. Health Commun. **23**(10–11), 909–955 (2018)
39. Brew-Sam, N., Chib, A.: How do smart device apps for diabetes self-management correspond with theoretical indicators of empowerment? An analysis of app features. Int. J. Technol. Assess. Health Care **35**(2), 150–159 (2019)
40. Caulfield, B.M., Donnelly, S.C.: What is connected health and why will it change your practice? QJM: Int. J. Med. **106**(8), 703–707
41. Huhn, S., Axt, M., Gunga, H.C., Maggioni, M.A., Munga, S., Obor, D., Sié, A., Boudo, V., Bunker, A., Sauerborn, R., Bärnighausen, T.: The impact of wearable technologies in health research: scoping review. JMIR Mhealth Uhealth **10**(1), e34384 (2022)
42. Tresp, V., Overhage, J.M., Bundschus, M., Rabizadeh, S., Fasching, P.A., Yu, S.: Going digital: a survey on digitalization and large-scale data analytics in healthcare. Proc. IEEE **104**(11), 2180–2206 (2016)
43. Lee, L., Egelman, S., Lee, J.H., Wagner, D.: Risk Perceptions for Wearable Devices. arXiv preprint: arXiv:1504.05694 (2015)
44. Sorber, J., Shin, M., Peterson, R., Cornelius, C., Mare, S., Prasad, A., Marois, Z., Smithayer, E., Kotz, D.: An amulet for trustworthy wearable mHealth. In: Proceedings of the Twelfth Workshop on Mobile Computing Systems & Applications, pp. 1–6 (2012)
45. Liu, J., Sun, W.: Smart attacks against intelligent wearables in people-centric internet of things. IEEE Commun. Mag. **54**(12), 44–49 (2016)
46. Perez, A.J., Zeadally, S.: Privacy issues and solutions for consumer wearables. It Prof. **20**(4), 46–56 (2017)
47. Sultan, N.: Reflective thoughts on the potential and challenges of wearable technology for healthcare provision and medical education. Int. J. Inf. Manage. **35**(5), 521–526 (2015)
48. Kim, W., Choi, J.: Challenges for wearable healthcare services. Int. J. Web Grid Serv. **12**(4), 407–417 (2016)
49. Loncar-Turukalo, T., Zdravevski, E., da Silva, J.M., Chouvarda, I., Trajkovik, V.: Literature on wearable technology for connected health: scoping review of research trends, advances, and barriers. J. Med. Internet Res. **21**(9), e14017 (2019)
50. Erkiliç, C.E., Yalçın, A.: Evaluation of the wearable technology market within the scope of digital health technologies. Gazi İktisat ve İşletme Dergisi **6**(3), 310–323 (2020)
51. Greiwe, J., Nyenhuis, S.M.: Wearable technology and how this can be implemented into clinical practice. Curr. Allergy Asthma Rep. **20**, 1–10 (2020)
52. Kerr, D., Butler-Henderson, K., Sahama, T.: Security, privacy, and ownership issues with the use of wearable health technologies. In: Cyber Law, Privacy, and Security: Concepts, Methodologies, Tools, and Applications, pp. 1629–1644. IGI Global (2019)
53. Siewiorek, D.: Wearable computing: retrospectives on the first decade. GetMobile: Mob. Comput. Commun. **21**(1), 5–10 (2017)
54. Motti, V.G.: Wearable health: opportunities and challenges. In: Proceedings of the 13th EAI International Conference on Pervasive Computing Technologies for Healthcare, pp. 356–359 (2019)
55. Genaro Motti, V., Caine, K.: An overview of wearable applications for healthcare: requirements and challenges. In: Adjunct Proceedings of the 2015 ACM International Joint Conference on Pervasive and Ubiquitous Computing and Proceedings of the 2015 ACM International Symposium on Wearable Computers, pp. 635–641 (2015)
56. Adams, D., Zheng, H., Sinclair, M., Murphy, M., McCullough, J.: Integrated care for pregnant women with type one diabetes using wearable technology. In: BIBE 2019; The Third

International Conference on Biological Information and Biomedical Engineering, pp. 1–5. VDE (2019)
57. Tanuwidjaja, E., Huynh, D., Koa, K., Nguyen, C., Shao, C., Torbett, P., Emmenegger, C., Weibel, N.: Chroma: a wearable augmented-reality solution for color blindness. In: Proceedings of the 2014 ACM International Joint Conference on Pervasive and Ubiquitous Computing, pp. 799–810 (2014)
58. Junker, H., Amft, O., Lukowicz, P., Tröster, G.: Gesture spotting with body-worn inertial sensors to detect user activities. Pattern Recogn. **41**(6), 2010–2024 (2008)
59. Kunze, K., Lukowicz, P.: Sensor placement variations in wearable activity recognition. IEEE Pervasive Comput. **13**(4), 32–41 (2014)
60. Nazneen, F., Boujarwah, F.A., Sadler, S., Mogus, A., Abowd, G.D., Arriaga, R.I.: Understanding the challenges and opportunities for richer descriptions of stereotypical behaviors of children with ASD: a concept exploration and validation. In: Proceedings of the 12th International ACM SIGACCESS Conference on Computers and Accessibility, pp. 67–74 (2010)
61. Lewy, H.: Wearable technologies–future challenges for implementation in healthcare services. Healthc. Technol. Lett. **2**(1), 2–5 (2015)
62. Wu, M., Luo, J.: Wearable technology applications in healthcare: a literature review. Online J. Nurs. Inform **23**(3) (2019)
63. Scully, C.G., Lee, J., Meyer, J., Gorbach, A.M., Granquist-Fraser, D., Mendelson, Y., Chon, K.H.: Physiological parameter monitoring from optical recordings with a mobile phone. IEEE Trans. Biomed. Eng. **59**(2), 303–306 (2011)
64. Soh, P.J., Vandenbosch, G.A., Mercuri, M., Schreurs, D.M.P.: Wearable wireless health monitoring: Current developments, challenges, and future trends. IEEE Microwave Mag. **16**(4), 55–70 (2015)
65. Wu, J., Li, H., Lin, Z., Goh, K.Y.: How big data and analytics reshape the wearable device market–the context of e-health. Int. J. Prod. Res. **55**(17), 5168–5182 (2017)
66. Scoble R, Israel S: Activity tracker. In: Wikipedia.com, Age of Context: Mobile, Sensors, Data and the Future of Privacy. https://en.wikipedia.org/wiki/Activity_tracker (2015)
67. The Hicon Team: Hicon smart wristband with social network icons. In: Indeegogo.com. https://www.indiegogo.com/projects/hicon-smartwristband-with-social-network-icons#/story (2015)
68. Withings: Pulse Ox Track Improve. In: Withings Inspire Health. 2009–2015. http://www.withings.com/eu/withings-pulse.html
69. Hu, H.Z., Feng, X.B., Shao, Z.W., Xie, M., Xu, S., Wu, X.H., Ye, Z.W.: Application and prospect of mixed reality technology in medical field. Curr. Med. Sci. **39**, 1–6 (2019)
70. Wang, Q., Markopoulos, P., Yu, B., Chen, W., Timmermans, A.: Interactive wearable systems for upper body rehabilitation: a systematic review. J. Neuroeng. Rehabil. **14**(1), 1–21 (2017)
71. Walmsley, C.P., Williams, S.A., Grisbrook, T., Elliott, C., Imms, C., Campbell, A.: Measurement of upper limb range of motion using wearable sensors: a systematic review. Sport. Med.-Open **4**, 1–22 (2018)
72. Liang, J., Xian, D., Liu, X., Fu, J., Zhang, X., Tang, B., Lei, J.: Usability study of mainstream wearable fitness devices: feature analysis and system usability scale evaluation. JMIR Mhealth Uhealth **6**(11), e11066 (2018)
73. Powell, L., Parker, J., Martyn St-James, M., Mawson, S.: The effectiveness of lower-limb wearable technology for improving activity and participation in adult stroke survivors: a systematic review. J. Med. Internet Res. **18**(10), e259 (2016)
74. Xu, J., Bao, T., Lee, U.H., Kinnaird, C., Carender, W., Huang, Y., Sienko, K.H., Shull, P.B.: Configurable, wearable sensing and vibrotactile feedback system for real-time postural balance and gait training: proof-of-concept. J. Neuroeng. Rehabil. **14**, 1–10 (2017)
75. Argent, R., Slevin, P., Bevilacqua, A., Neligan, M., Daly, A., Caulfield, B.: Clinician perceptions of a prototype wearable exercise biofeedback system for orthopaedic rehabilitation: a qualitative exploration. BMJ Open **8**(10), e026326 (2018)

76. Hou, Z., Liu, X., Tian, M., Zhang, X., Qu, L., Fan, T., Miao, J.: Smart fibers and textiles for emerging clothe-based wearable electronics: materials, fabrications and applications. J. Mater. Chem. A **11**(33), 17336–17372 (2023)
77. Anusha, A.S., Jose, J., Preejith, S.P., Jayaraj, J., Mohanasankar, S.: Physiological signal based work stress detection using unobtrusive sensors. Biomed. Phys. Eng. Express **4**(6), 065001 (2018)
78. Arza, A., Garzón-Rey, J.M., Lázaro, J., Gil, E., Lopez-Anton, R., de la Camara, C., Laguna, P., Bailon, R., Aguiló, J.: Measuring acute stress response through physiological signals: towards a quantitative assessment of stress. Med. Biol. Eng. Compu. **57**, 271–287 (2019)
79. Chalmers, T., Hickey, B.A., Newton, P., Lin, C.T., Sibbritt, D., McLachlan, C.S., Clifton-Bligh, R., Morley, J., Lal, S.: Stress watch: the use of heart rate and heart rate variability to detect stress: a pilot study using smart watch wearables. Sensors **22**(1), 151 (2021)
80. Herborn, K.A., Graves, J.L., Jerem, P., Evans, N.P., Nager, R., McCafferty, D.J., McKeegan, D.E.: Skin temperature reveals the intensity of acute stress. Physiol. Behav. **152**, 225–230 (2015)
81. Mozos, O.M., Sandulescu, V., Andrews, S., Ellis, D., Bellotto, N., Dobrescu, R., Ferrandez, J.M.: Stress detection using wearable physiological and sociometric sensors. Int. J. Neural Syst. **27**(02), 1650041 (2017)
82. Nath, R.K., Thapliyal, H., Caban-Holt, A.: Machine learning based stress monitoring in older adults using wearable sensors and cortisol as stress biomarker. J. Signal Process. Syst. 1–13 (2021)
83. Sandulescu, V., Andrews, S., Ellis, D., Bellotto, N., Mozos, O.M.: Stress detection using wearable physiological sensors. In: Artificial Computation in Biology and Medicine: International Work-Conference on the Interplay Between Natural and Artificial Computation, IWINAC 2015, Elche, Spain, June 1–5, 2015, Proceedings, Part I 6, pp. 526–532. Springer International Publishing (2015)
84. Egger, S.T., Knorr, M., Bobes, J., Bernstein, A., Seifritz, E., Vetter, S.: Real-time assessment of stress and stress response using digital phenotyping: a study protocol. Front. Digit. Health **2**, 544418 (2020)
85. Ozkan-Yildirim, S., Pancar, T.: Smart wearable technology for health tracking: what are the factors that affect their use? IoT Healthc. Ambient. Assist. Living 165–199 (2021)
86. Kim, T.B., Ho, C.T.B.: Validating the moderating role of age in multi-perspective acceptance model of wearable healthcare technology. Telemat. Inform. **61**, 101603 (2021)
87. Due, B.L.: The Future of Smart Glasses: An Essay About Challenges and Possibilities with Smart Glasses, vol. 1, pp. 1–21. København, Denmark: Centre of Interaction Research and Communication Design, University of Copenhagen (2014)
88. Kim, K.J., Shin, D.H.: An acceptance model for smart watches: Implications for the adoption of future wearable technology. Internet Res. **25**(4), 527–541 (2015)
89. Lee, J., Kim, D., Ryoo, H.Y., Shin, B.S.: Sustainable wearables: wearable technology for enhancing the quality of human life. Sustainability **8**(5), 466 (2016)
90. Knight, H.M., Gajendragadkar, P.R., Bokhari, A.: Wearable technology: using Google Glass as a teaching tool. Case Rep. **2015**, bcr2014208768 (2015)
91. Signorini, M.G., Fanelli, A. and Magenes, G.: Monitoring fetal heart rate during pregnancy: contributions from advanced signal processing and wearable technology. Comput. Math. Methods Med. (2014)
92. Choi, J., Kim, S.: Is the smartwatch an IT product or a fashion product? A study on factors affecting the intention to use smartwatches. Comput. Hum. Behav. **63**, 777–786 (2016)
93. Zhang, M., Luo, M., Nie, R., Zhang, Y.: Technical attributes, health attribute, consumer attributes and their roles in adoption intention of healthcare wearable technology. Int. J. Med. Informatics **108**, 97–109 (2017)
94. Tehrani, K., Michael, A.: Wearable technology and wearable devices: everything you need to know. Wearable Devices Mag. **26** (2014)
95. Li, H., Wu, J., Gao, Y., Shi, Y.: Examining individuals' adoption of healthcare wearable devices: an empirical study from privacy calculus perspective. Int. J. Med. Informatics **88**, 8–17 (2016)

96. Chan, M., Estève, D., Fourniols, J.Y., Escriba, C., Campo, E.: Smart wearable systems: current status and future challenges. Artif. Intell. Med. **56**(3), 137–156 (2012)
97. Chau, K.Y., Lam, M.H.S., Cheung, M.L., Tso, E.K.H., Flint, S.W., Broom, D.R., Tse, G., Lee, K.Y.: Smart technology for healthcare: exploring the antecedents of adoption intention of healthcare wearable technology. Health Psychol. Res. **7**(1) (2019)
98. Park, E., Kim, K.J., Kwon, S.J.: Understanding the emergence of wearable devices as next-generation tools for health communication. Inf. Technol. People **29**(4), 717–732 (2016)
99. Lamb, K., Huang, H.Y., Marturano, A., Bashir, M.: Users' privacy perceptions about wearable technology: examining influence of personality, trust, and usability. In: Advances in Human Factors in Cybersecurity: Proceedings of the AHFE 2016 International Conference on Human Factors in Cybersecurity, July 27–31, 2016, Walt Disney World®, Florida, USA, pp. 55–68. Springer International Publishing (2016)
100. Meier, D.Y.: Influence of Cultural Factors on Wearable Technology Acceptance in Healthcare: An Empirical Study with Chinese and Swiss Consumers (2019)
101. Gay, V., Leijdekkers, P.: Bringing health and fitness data together for connected health care: mobile apps as enablers of interoperability. J. Med. Internet Res. **17**(11), e260 (2015)
102. Kalantari, M.: Consumers' adoption of wearable technologies: literature review, synthesis, and future research agenda. Int. J. Technol. Mark. **12**(3), 274–307 (2017)

AI and IoT Enabled Healthcare Applications

N. Krishna Chaitanya, K. V. Prasada Reddy, V. S. V. Harika, and G. Kameswari

Abstract This chapter discusses about the various technologies and components used in the healthcare using Artificial Intelligence (AI) and the Internet of Things (IoT). The role is to detect the disease with better accuracy. To tackle the healthcare system in an efficient way it is better to use the technology. The preferred technologies that help the healthcare system are IoT and artificial intelligence. IoT provides a platform that helps to share information through the communication network. It helps to gather various parameters of the patient and to monitor the health condition. The key features of IoT in the healthcare profession will be thoroughly discussed. AI is helpful in the diagnosis and remedial services to the patients. We are sure that this chapter will bring an attention of the people who are working in the healthcare centers where innovation plays a crucial role. This technology is going to fill the gap between the patient and doctor under critical conditions.

Keywords AI · IoT · Healthcare · Patient · Doctor · Technology

1 Introduction

In today's human life internet has become a part of it. At the early stages of internet, it has been used for the sake of communication among the people in the form of mails, voice calls, video calls, etc. With the advancements in technology, now most of the electronic devices have become very smart. The device which works with the

N. Krishna Chaitanya (✉)
Department of ECE, Sree Venkateswara College of Engineering, Nellore, Andhra Pradesh 524316, India
e-mail: nosinakc@gmail.com

K. V. Prasada Reddy · V. S. V. Harika
Department of AI & DS, Sree Venkateswara College of Engineering, Nellore, Andhra Pradesh 524316, India

G. Kameswari
Department of ECE, NBKRIST, Vidyanagar, Tirupati, Andhra Pradesh 524413, India

S. Mahajan et al. (eds.), *Smart Systems: Engineering and Managing Information for Future Success*, Information Systems Engineering and Management 22,
https://doi.org/10.1007/978-3-031-76152-2_5

internet are popularly called as smart devices. Now the devices which are connected through the internet is operated from anywhere without any boundaries towards the operating location. The basic architecture [1] for IoT is shown in Fig. 1.

From Fig. 1, sensors are the devices which collect data and send via internet gateway to reach to the cloud. Gateway is the device which acts as an intermediate node between communication network and sensor network. Communication network provides real time information sharing. But, a reliable and secured connection has to be ensured. The cloud is a storage device where the user data is saved in service provider memory. Then the data can be accessed by the intended people, who have the authorization over the cloud. With the rapid growth in technology, such as manufacturing smart devices and the availability of high data rates in the internet, IoT has changed its perception to real world applications. Now most of the global fields are dependent on the internet. The usage of the internet has lifted out the boundaries from mail usage to unlimited applications. At the earlier stages of IoT, researchers are targeted to operate few devices like, fan, air-conditioner and geyser. Now IoT is covering almost all the areas of research as mentioned below.

IoT finds key applications in the following areas [2] of

1. Smart home
2. Agriculture
3. Industry
4. Transport
5. Education
6. Consumer
7. Manufacturing
8. Military
9. The environment monitoring
10. Energy management
11. Healthcare.

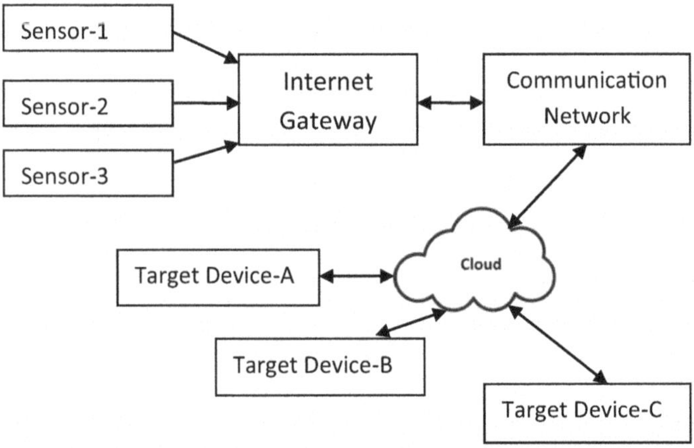

Fig. 1 Basic architecture of IoT

Healthcare is the predominant application of IoT [3]. In the next section the main focus is on IoT in healthcare applications.

2 Literature Review

One of the most important fields in any country is healthcare. Most of the healthcare systems are traditional, where the patients visit the hospital. Based on patient symptoms, the appropriate tests are done for the diagnosis of the decease. If the basic evaluation parameters of a patient such as heartbeat, blood-glucose, and blood pressure are measured with the help of traditional methods such as the stethoscope, one test glucose meter and BP apparatus. These tests are done when the patient feels sick and sometimes patient health condition becomes critical. Under critical conditions, there is a scope to lose the life of the patient because of not detecting the symptoms early.

Chao His Huang proposed [4] a system which uses 2G network, sensors, bluetooth, and zigbee to transfer the data. Here the drugs are supplied with the system for the patient.

Boyi and his team proposed [5] a method uses cloud computing that is an advanced system for patient health monitoring. This method is cost effective.

Avik Ghose et al., proposed [6] a system for monitoring the health condition of elderly people. This system just monitors the patient condition only and decisions are not made by system.

3 IoT in Healthcare

One of the most important areas in human life is health. Most of the governments give the priority to healthcare and education. Due to food habits, climatic conditions, people are facing a lot of health issues.

If any person has been admitted due to severe health issues, then the patient will be physically monitored. Based on the diagnosis done, an appropriate treatment is done to the patient. With technology growth, patient monitoring [7, 8] has become an easy task. Where the sensors are connected to the patient and the corresponding data is fetched from the data. The data is processed and if any adverse effects are identified, immediate alert will be given to nurses and doctors, where quick response can be expected to save the patient life.

The major advantages of IoT in the healthcare

a. Reduction in cost:

   This IoT devices used by the patient, it enables the monitoring the patient in real time, thereby it reduces the patient stay in the hospital, and also reduces frequent visits of the doctor during the stay at the hospital.

b. Faster Diagnosis of disease:

   With patient, real-time health condition data, it is easy to diagnose the disease at the early stage.

c. Better Treatment:

   Because of the early detection, it is easy for the doctors to give a better treatment.

d. Capable of reducing human errors:

   With the help of IoT devices, it is possible to get the effective data and there by avoids the possible of misreading the patient data.

e. Continuous observation:

   With the help of this IoT, it is possible to have regular observation of the patient condition without physical appointment.

3.1 IoT Applications in Healthcare

The most popular IoT applications are

1. Monitoring of Blood Glucose
2. Monitoring of Blood Pressure
3. Monitoring of Heart Rate
4. Monitoring of Temperature
5. Monitoring of Depression
6. Inhalers for Asthma
7. Connected Contact Lenses
8. Robotic Surgery.

3.2 Blood Glucose Monitoring

If the blood glucose levels are above normal limits it causes a Diabetes Mellitus (DM) metabolic disorder [9]. Lack of enough insulin in the body will induce diabetes. Insulin basically controls the glucose levels in blood. As per WHO data more than 500 million people are suffering from diabetes.

There are two types of diabetes. First one is type-1, which appears commonly in teenagers, adults, as well as in children. Type-1 is due to loss of insulin. Type-2 is commonly appears in the elder persons due to defect in insulin secretion. This diabetes can be monitored with the help of Invasive, noninvasive and minimally invasive. With the advancements in technology sensors are available; it has become easy to monitor, the glucose levels continuously. The sensor that is used to measure the glucose levels is shown in Fig. 2. The complete setup for getting the patient glucose levels from the sensor to the doctor phone is shown in Fig. 3.

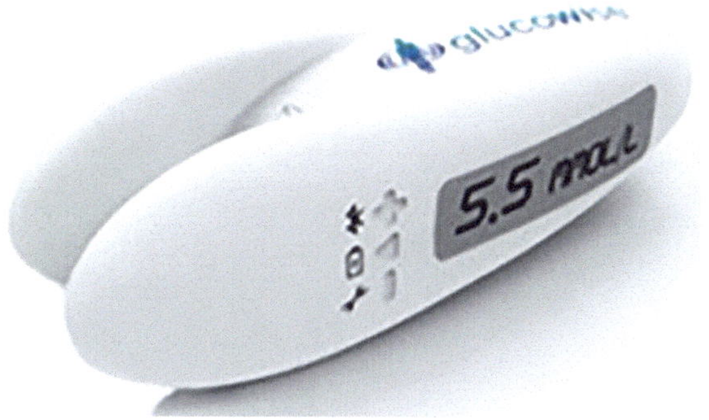

Fig. 2 A sensor to measure glucose levels

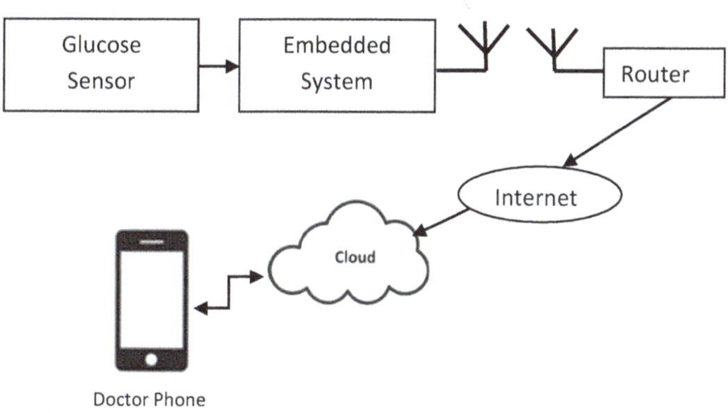

Fig. 3 Setup for getting the patient data from the sensor to the doctor

3.3 Blood Pressure Monitoring

The important parameter that is used to indicate the physiological information of a human body is blood pressure (BP) [10]. A number of devices are developed to measure blood pressure. The continuous measurement of BP is desirable for the people at home, or healthcare. Various techniques of non-invasive are used to measure blood pressure. These techniques are continuous and non-continuous. Non-continuous methods like oscillometric as well as auscultatory, and in continuous methods like plethysmography, arterial tonometry, and pulse transit time.

Wireless blood pressure devices are available in the market and one such device is shown in Fig. 4 [11]. This is a high accuracy blood pressure sensor that is wearable to arm. This reads the data and sends to the cloud through wireless modem. Finally, the data can be seen in mobile (Fig. 5).

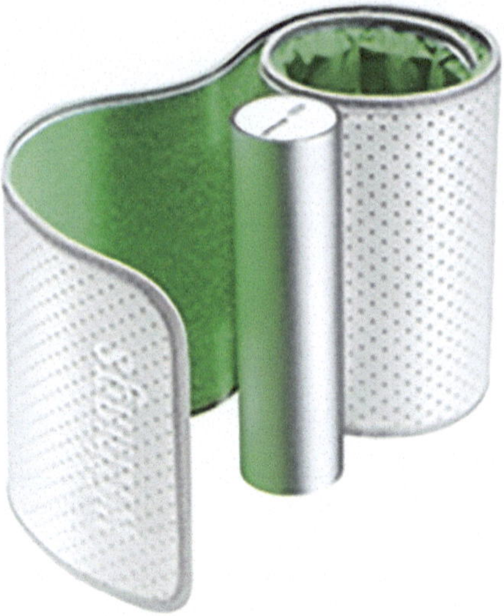

Fig. 4 Wearable blood pressure sensor

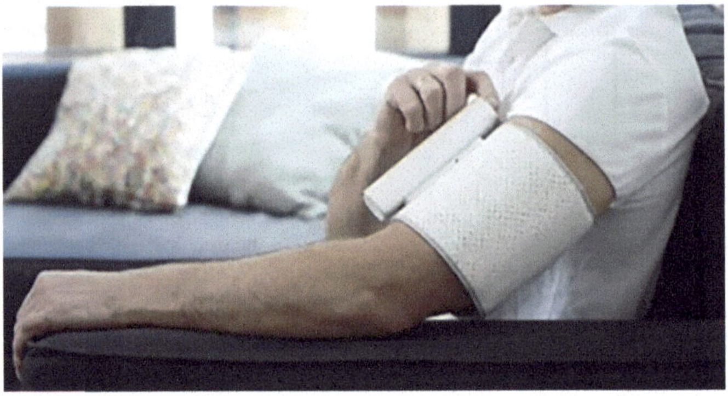

Fig. 5 Blood pressure sensor fixed to arm

3.4 Heart Rate Monitoring

These days, cardiac arrest has become a major concern [12]. With the advancement in technology, wearable smart devices are connected with IoT that provides a solution for heart patients. In the olden days cardiac arrest has seen only in the elder people. In the present generation, it has been seen in teenagers also. The resting heart rate in the children is who are below the age of 17 is in the range of 65 and 115 beats per minute (bpm). In case of adults, who ages from 17 to 60 the resting heart rate in the range of 60 and 100 bpm. In the elderly people, who are aged above 60 years, their heartbeat rate in the range of 65 and 120 bpm. The wireless heart rate monitoring

Fig. 6 The wireless heart rate monitoring device [13]

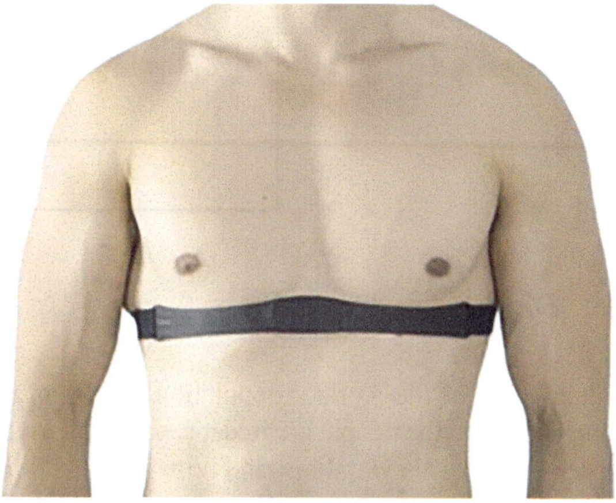

Fig. 7 Usage of wireless heart rate sensor

device is shown in Fig. 6 and how the device is fixed as a sensor in our body is shown in Fig. 7.

3.5 Temperature Monitoring

The normal body temperature varies from person to person based on the age. The accepted body temperature is in the range from 97 °F (36.1 °C) to 99 °F (37.2 °C). To measure the temperature and to send it wirelessly a number of sensors available. One of the most useful sensors which are in the form of a patch is shown in Fig. 8. It is capable to share the real time data to the server through the internet.

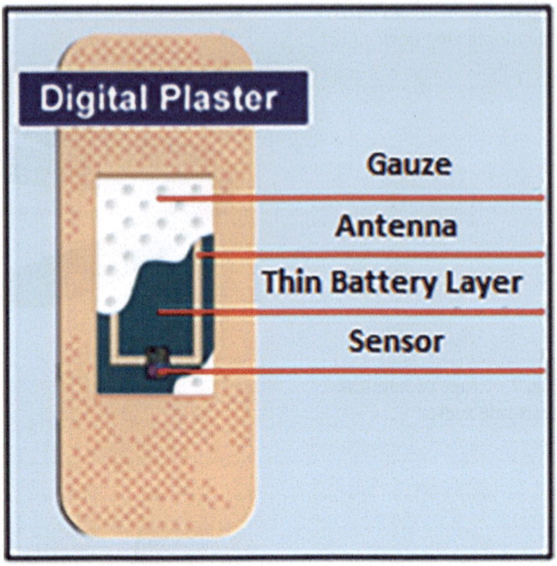

Fig. 8 A wireless plaster [14] for temperature monitoring

3.6 Depression Monitoring

Around 25% of the world's population has been affected by depression [15–17] annually. When the people are in depression experiencing psychological difficulties as well as physical symptoms, and it may also leading suicides at higher rates. As a result, depression has become an important public health issue. But identifying the depression symptoms is not straightforward. To find the depression stage, healthcare professionals go through a phase of conducting interviews as well as self-report questionnaires. But regrettably, these methods take a lot of time, pricey, and even professionals involvement is also required. These can be eliminated with the help of smart devices and the complete setup for depression monitoring is shown in Fig. 9.

With the help of above setup, daily activity of the patient is identified. Based on the reports generated from it, appropriate action will be taken. It also saves the patients who are in critical condition and may take serious action such as suicides.

3.7 Asthma Inhalers

Asthma is a respiratory disease [18], which is similar to chronic obstructive pulmonary disease (COPD). It factors such as smoking, air-pollution, experience to convinced chemicals or perfumes, etc. As per data of the World Health Organization (WHO), COPD has become the top leading source of deaths worldwide at the time of 2030. This problem has driven the market for brining devices of digital

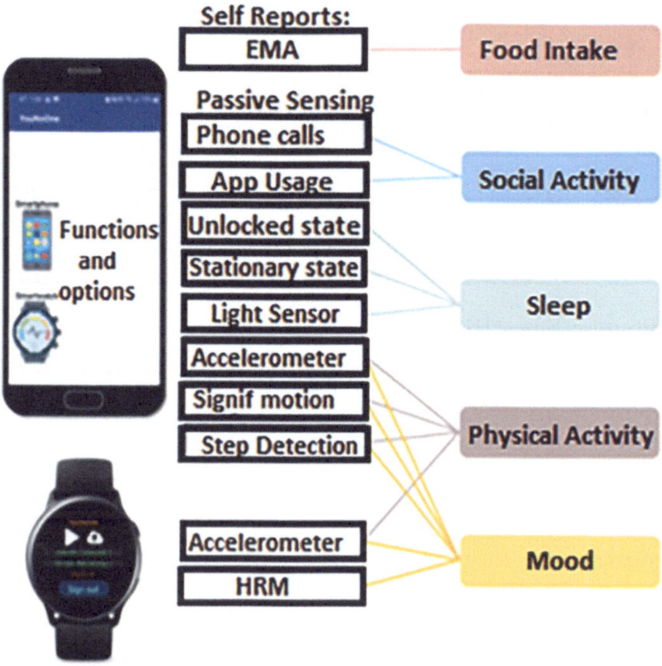

Fig. 9 Depression monitoring equipment

respiratory, which take in both therapeutic (inhalers, nebulizers) and devices for diagnostic (Fig. 10).

Fig. 10 Smart asthma inhaler [19]

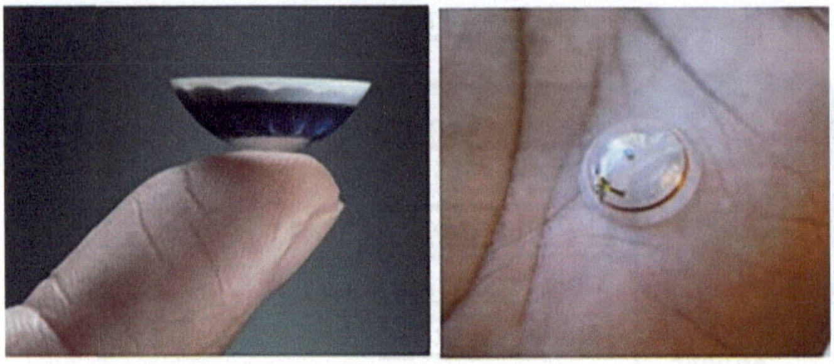

Fig. 11 Smart contact lenses [21]

The purpose of this smart inhaler is to help patients and their doctors monitor adherence and "other treatment factors".

3.8 Connected Contact Lenses

It [20] is an interface among the human body and the electronic device in wearable healthcare applications. Despite over investigations about smart contact lenses used in diagnostic applications, there is no availability of drug delivery-control mechanism along with the analysis of real-time biometric. The smart contact lens is shown in Fig. 11. This is mainly used to detect glucose levels. It is also going to find an application in smart vision.

3.9 Robotic Surgery

Robots are become a focal point in the healthcare as well as in medical industries, where they are performing their duty using electronics and mechanics [22].

The duties that are done by robots are

- Caring of patients
- Rehabilitation
- Artificial prosthetics
- Medical interventions
- E-health, monitoring.

The involvement of robots in surgery is shown in Fig. 12. Robotics along with healthcare automation allowing the use of interconnected robots that are used in the healthcare which is based on IoT. An IoT platform provides a number of facilities through interconnected objects or people. This environment is used for transferring

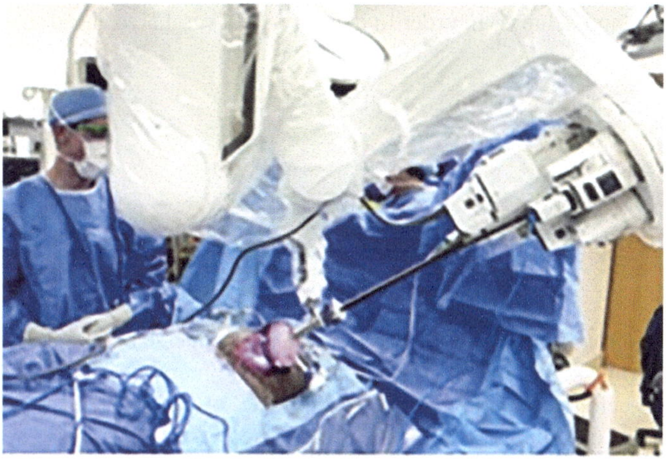

Fig. 12 Robotic surgery

the data among the robots—without human intervention. With the support of humans, robots are capable to perform advanced surgeries.

Though there are numerous applications available, at the same time there are few challenges [23] in the IoT in healthcare. The challenges are

1. Reliability of the Internet: Make sure that the connection is always there between the sensor device and the target device (doctors).
2. Device to device connectivity: It is very important that the connection from one stage to another stage properly connected and communication.
3. Resource Management: resources like data storage, processing and their corresponding devices.
4. Data processing: How the data is transmitted from the sensors and gets stored in the cloud and the data is being processed for further action is very important.
5. Security: To provide security, then it is very important to use the encryption and decryption techniques.

4 Role of Artificial Intelligence in Healthcare

The most trending research area is "Artificial Intelligence (AI)". It has made the machines to think and act like humans. Now a day most of the research scholars, scientists, academicians, and industry people are working together for a tremendous change in artificial intelligence. This is mainly possible by knowing, how humans are thinking, behaving, learning, and solving problems. In order to make the machines to think like humans, it is necessary to develop intelligent software as well as systems. The goal of AI is to think, act, and behave like humans. In fact, it's a difficult process to understand and achieve. In order to develop AI, several areas need to contribute and are

- Computer Science
- Maths
- Neuron science
- Biology
- Sociology
- Psychology
- Philosophy.

In AI, along with major areas, it is necessary to understand about intelligence. It has the ability to store and retrieve data from the memory, language speaking, and writing, seeing, calculating, reacting, emotions, solving problems, getting ideas, thinking and so on. Mainly intelligence is composed of

- Learning
- Reasoning
- Problem solving
- Perception
- Linguistic intelligence.

The major difference between human intelligence and machine intelligence is that humans are able to store and recall information with the help of patterns, whereas machines with a set of rules and data.

4.1 Applications of AI

Most popular applications of AI is in

- Healthcare
- Data security
- Retail
- E-commerce
- Chatbots
- Robotics
- Education
- Transport
- Agriculture
- Finance
- Gaming
- Astronomy
- Entertainment
- Automotive
- Social Media.

Here the discussion is confined to the application of healthcare. AI will help to diagnosis the decease faster than human with accuracy [16, 17]. Based on the results, the system will help the patient to go for early treatment by giving a warning. This is possible with artificial intelligence. Artificial intelligence has found major applications [24] in the following areas

- Diagnostic imaging
- Genetic
- Electro diagnosis
- Disability evaluation
- Nervous
- Cardio vascular
- Digestive
- Skin.

In order to detect the decease, artificial intelligence uses a various machine learning algorithms such as,

- Support vector machines
- Neural Networks
- Hidden Markov
- Linear Regression
- Logistic regression
- Random forest
- Nearest neighbor
- Decision tree.

With the help of the above machine learning algorithms as shown in Fig. 13, AI device analyses data such as genetic, imaging and other information. With the help of natural language processing, it extracts the information such as clinical notes. The most widely used machine learning algorithm is the support vector machine.

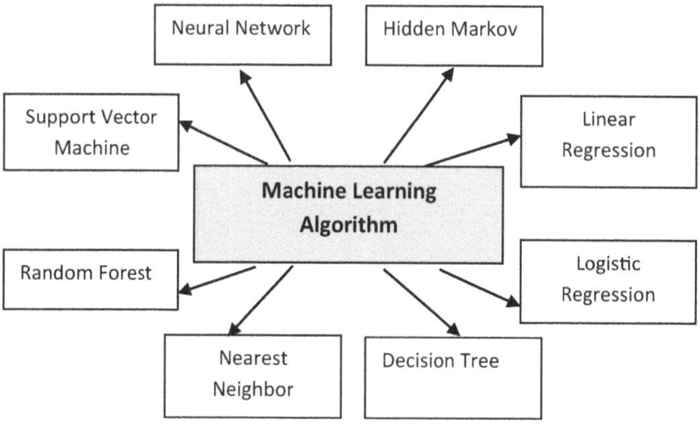

Fig. 13 The most widely used machine learning algorithms used in AI

The reasons go ahead with artificial intelligence is, it is very difficult for the administration in healthcare and clinical support. In case of healthcare administration, if a patient comes to the hospital, then the data will be entered into the system. If the patient is repeatedly comes to the hospital, then the complete data will be available in the system. With the help of AI, it is very useful for clinicians to access the patient data and to track for the patient condition. Even it is possible to accurately detect the decease with the help of advanced machine learning algorithms. It helps to take decisions in clinical reports. Systems with pre loaded clinical data and knowledge is able to reduce medical errors and to improve the efficiency and consistency of clinical reports for healthcare professionals [23]. This AI system in medical reports is being used from 1970s. With the advancements in AI systems, it has become very easy and effective to diagnose the decease and to warn the patient at earlier stages, thereby it saves the lives of the patients. With AI, patient health will be monitored such as sleeping patterns, blood pressure, heart rate, etc. AI really makes use of the machine learning methods for the use of personalized medicine, patient monitoring, drug discovery, patient record management.

Artificial Intelligence is combined with IoT which is shown in Fig. 14. AI is basically used for diagnosis and to give accurate results of the disease. There by it becomes very easy for the doctors to understand the current position of the patient and to give a better treatment. This is possible with the pre loaded data about all the diseases.

The role of AI is to collect data from the patient sensors and does the processing using machine learning and natural language processing. The data is stored in the cloud permanently. Based on the patient data, it will generate the disease and the

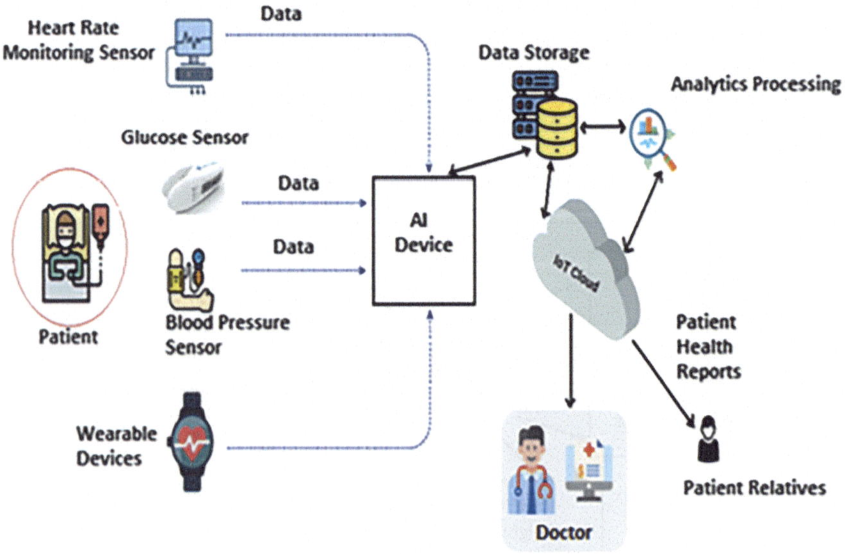

Fig. 14 AI with IoT

current level of criticality. If IoT gets added to AI, it gives a real time alert to the doctors about the patient condition as well as to the patient's relatives. For example, if a heart beat sensor gets connected to the patient, if there is any abnormality in the heartbeat and it will be detected by AI. Then alert will be sent to the doctor as well as patient relatives with the help of IoT. Thereby, it becomes easy for the doctor to give appropriate treatment in right time.

In addition to technological developments, still there are challenges associated.

1. Retrospective versus prospective studies
2. Peer reviewed randomized control trails as an evidence gold standard
3. Metrics often doesn't reflect clinical applicability
4. Difficulty in comparing different algorithms
5. Challenges related to machine learning science
6. Dataset shift
7. Accidentally fitting confounders versus true signal
8. Algorithmic bias.

5 Challenges in IoT in Healthcare

The major challenges of IoT in the healthcare system are

1. Proto type Development
2. Authentication
3. Robustness and flexibility
4. Data transmission
5. Cost
6. Connectivity
7. Data Storage
8. Privacy
9. Energy requirements
10. Integrity.

5.1 Proto Type Development

It is a difficult task to develop a prototype that suits to the intended application with the necessary sensors. Based on the structure of interconnection, it will decide the effective communication between the sensor information along with the simulation.

5.2 Authentication

It is necessary to provide authenticate in order to verify the information that has been stored for a patient.

5.3 Robustness and Flexibility

The system has to be as robust as it is to provide a steady behavior of the intended system. At the system time the system has to be flexible for the real time changes in the architecture and communication among the devices.

5.4 Data Transmission

All the sensors read the information and transmit to the central control unit. In order to transmit the data to the control unit, it is necessary to ensure it has been received without any sort of noise addition.

5.5 Cost

Due to the increased number of sensors, that requires high speed internet, good controller along with an efficient algorithm, the implementation cost also increases.

5.6 Connectivity

Here the entire communication among the sensors and controller is based on reliable connectivity. If there is a connection lost, chances for the loss of life.

5.7 Data Storage

In order to store the information received from the sensors, it requires large storage capacity. Under these conditions one option is using the cloud services.

5.8 *Privacy*

There are more chances for the IoT devices for attack in the IoT environment.

5.9 *Energy Requirements*

If more number of devices is connected to the environment, devices consume more amount of energy.

5.10 *Integrity*

It is very important that, how the sensors and other equipment is connected for effective communication among them without any disturbances.

6 Conclusions

The Internet of things (IoT) is going to become a major support in healthcare in the future. Here, it can be concluded that the IoT in healthcare is trying to fill the gap between patient and doctors. At the same time, effective and efficient monitoring of the patient is possible with the technology, where it may not be possible with the human involvement. Diagnoses also become accurate and it is easy to identify the decease. The future is going to be expected with the smart IoT devices that takes cares of patient life. In Addition with this AI plays a crucial role in detecting the disease and to understand the level of patient criticality. Combining both the technologies together to get better accuracy to detect and give treatment. In the next few years, we can expect the use of AI with the IoT in healthcare.

References

1. Alam, M.M., et al.: A survey on the roles of communication technologies in IoT-based personalized healthcare applications. IEEE Access **6**, 36611–36631 (2018)
2. Ramson, S.R.J., Vishnu, S., Shanmugam, M.: Applications of Internet of things (IoT)—an overview. In: 2020 5th International Conference on Devices, Circuits and Systems (ICDCS). IEEE (2020)
3. Carnaz, G., Nogueira, V.B.: An Overview of IoT and Healthcare. Escola de Ciências e Tecnologia da Universidade de Évora (2016)
4. Huang, C.H., Cheng, K.W.: RFID technology combined with IoT application in medical nursing system. Bull. Netw. Comput. Syst. Softw. **3**(1), 20–24 (2014)

5. Xu, B., Xu, L., Cai, H., Jiang, L., Luo, Y., Gu, Y.: The Design of an m-Health Monitoring System Based on a Cloud Computing Platform. Talor & Francis (2015)
6. Ghose, A., et al.: UbiHeld: ubiquitous healthcare monitoring system for elderly and chronic patients. In: Proceedings of the ACM Conference on Pervasive and Ubiquitous Computing Adjunct Publication, pp. 8–12. Zurich, Switzerland (2013)
7. Goyal, S., et al.: IoT enabled technology in secured healthcare: applications, challenges and future directions. In: Cognitive Internet of Medical Things for Smart Healthcare, pp. 25–48. Springer, Cham (2021)
8. Kumar, S.A., Mahesh, G.: IoT in smart healthcare system. In: Internet of Things for Healthcare Technologies, pp. 1–19. Springer, Singapore (2021)
9. Valenzuela, F., et al.: An IoT-based glucose monitoring algorithm to prevent diabetes complications. Appl. Sci. **10**(3), 921 (2020)
10. Dinh, A., Luu, L., Cao, T.: Blood pressure measurement using finger ECG and photoplethysmogram for IoT. In: International Conference on the Development of Biomedical Engineering in Vietnam. Springer, Singapore (2017)
11. https://iot.do/devices/withings-wireless-blood-pressure-monitor
12. Abba, S., Garba, A.M.: An IoT-based smart framework for a human heartbeat rate monitoring and control system. Multi. Digit. Publ. Inst. Proc. **42**(1) (2019)
13. https://in.pinterest.com/pin/567242515552886118/
14. Reddy, S., Fox, J., Purohit, M.P.: Artificial intelligence-enabled healthcare delivery. J. R. Soc. Med. **112**(1), 22–28 (2019)
15. Narziev, N., et al.: STDD: short-term depression detection with passive sensing. Sensors **20**(5), 1396 (2020)
16. Krishna Chaitanya, N., et al.: IoT-enabled moving wheelchair with obstacle detection and continuous health monitoring. In: 2024 Second International Conference on Emerging Trends in Information Technology and Engineering (ICETITE). IEEE (2024)
17. Krishna Chaitanya, N., Swarna Latha, T.: Introduction to cybersecurity with AI, ML, and blockchain. In: Kaushik, K., Sharma, I. (eds.) Next-Generation Cybersecurity. Blockchain Technologies. Springer, Singapore (2024). https://doi.org/10.1007/978-981-97-1249-6_1
18. https://www.electronicsforu.com/market-verticals/medical/smart-inhalers-enabling-better-respiratory-care
19. https://www.asthma.org.uk/advice/inhalers-medicines-treatments/inhalers-and-spacers/smart-inhalers/
20. Kim, S.-K., et al.: Wireless smart contact lens for diabetic diagnosis and therapy. Sci. Adv. **6**, 17 (2020)
21. https://www.wired.com/story/mojo-vision-smart-contact-lens/
22. Patel, A.R., et al.: Vitality of robotics in healthcare industry: an Internet of things (IoT) perspective. In: Internet of Things and Big Data Technologies for Next Generation Healthcare, pp. 91–109. Springer, Cham (2017)
23. Dang, L.M., et al.: A survey on internet of things and cloud computing for healthcare. Electronics **8**(7), 768 (2019)
24. Jiang, F., et al.: Artificial intelligence in healthcare: past, present and future. Stroke Vasc. Neurol. **2**, 4 (2017)

Exploring AI's Prowess in Advancing Cybersecurity

Ananyaa Sharma, Riya Bhatia, Dolly Sharma, and Ashima Kalra

Abstract Cybersecurity issues are always changing as the digital landscape grows, making creative solutions to protect sensitive data and important assets necessary. With its sophisticated capabilities for threat detection, reaction automation, and wise decision-making, AI has become a powerful tool in the battle against cyber threats. When it comes to cybersecurity, information, and operational technology (OT) security are closely interconnected practices, technologies, and concepts. One of the key aspects of cybersecurity is to offensively conduct penetration tests, also known as ethical hacking and launching cyberattacks such as distributed denial of service (DDoS) attacks, data breaches, or malware deployment to combat adversaries. The integration of AI in cybersecurity plays a critical role. It has the capability to identify risks, vulnerabilities, and irregularities within your network or system. By leveraging AI, you can uncover patterns that indicate a potential breach. Moreover, AI assists in spotting risks, weaknesses, and anomalies, thereby strengthening your network and system security. This chapter delves into the ways in which AI is advancing cybersecurity. It harnesses machine learning, deep learning, and other AI techniques to improve the accuracy of threat detection, automate security operations, and aid security professionals in effectively mitigating and responding to cyber threats.

Keywords Cybersecurity · AI · Cyber threats · Vulnerability · Machine learning

A. Sharma · R. Bhatia · D. Sharma (✉)
Department of Computer Science and Engineering, Amity University, Noida, India
e-mail: dolly.azure@gmail.com

A. Kalra
Department of Electronics and Communication Engineering, Chandigarh Engineering College-CGC, Landran, Mohali, India

S. Mahajan et al. (eds.), *Smart Systems: Engineering and Managing Information for Future Success*, Information Systems Engineering and Management 22,
https://doi.org/10.1007/978-3-031-76152-2_6

1 Intersection of AI and Cybersecurity

Cybercriminals are growing more skilled and outpacing existing cybersecurity measures as technology develops daily. To proactively combat emerging cyberattacks and outsmart cybercriminals, experts are now incorporating Artificial Intelligence (AI) into their defence strategies. AI, a field of computer science, employs complex mathematical algorithms to emulate human thought processes. The term "artificial intelligence" was coined by John McCarthy and other scholars in 1956. The initial advancements in AI were facilitated through games like checkers, which machines could learn and improve through practice, surpassing the abilities of average players. This marked a significant milestone for AI within the realm of digital computing, marking the end of its first decade [1, 2]. The development of AI throughout its history is evident, as enterprises now have the capability to design, manage, and deploy machine learning and deep learning models at scale with the assistance of AI platforms. These platforms help make AI technology more accessible and cost-effective by simplifying tasks such as data management and deployment, alleviating the software development burden.

As a response to the growing threat of cybercrime, the application of artificial intelligence (AI) in monitoring and preventing cyber threats is expanding on a larger scale. AI technologies, such as machine learning and predictive analytics, enhance cybersecurity measures by automating threat detection, response, and prevention. Machine learning algorithms can analyse vast datasets to identify patterns, anomalies, and potential security breaches in real time, providing a proactive defence against a wide range of cyber threats. Additionally, AI enables the development of adaptive security systems capable of learning and evolving to counter new and sophisticated attack vectors. While AI enhances the efficacy of cybersecurity measures, it also introduces challenges, such as the need to defend against adversarial attacks aiming to exploit vulnerabilities in AI systems. Striking a balance between innovation and security is crucial as organizations leverage AI to strengthen their cybersecurity posture in the face of an ever-evolving digital threat landscape. Cybersecurity is a linchpin for the functioning of our modern, digitally driven society. Its importance extends beyond individual privacy to encompass economic stability, national security, and the overall resilience of our interconnected global community.

1.1 The Backstory

As businesses and individuals face an increasing number of complex cyberthreats in the modern day, cybersecurity has grown in significance. Many industries are undergoing digital transformation to improve efficiency, productivity, and competitiveness. This transformation involves the integration of digital technologies into all aspects of operations, including manufacturing, supply chain management, energy production, and transportation. As a result, traditional OT systems are being modernized and interconnected with IT systems to enable real-time data collection, analysis, and decision-making. Traditional cybersecurity approaches, although effective, often struggle to keep pace with the evolving tactics and strategies employed by malicious actors. IT systems have long been the primary target for cyber-attacks, with threats ranging from malware and ransomware to phishing and insider threats. However, OT systems are now becoming increasingly targeted by cyber adversaries seeking to disrupt critical infrastructure, cause physical harm, or steal sensitive information. Cybersecurity professionals must understand the unique security risks and threat landscape associated with both IT and OT environments and develop tailored defense strategies to mitigate these risks effectively [3].

Connected healthcare systems incorporate a wide range of medical devices and sensors that collect patient data in real-time. These OT components include wearable devices, implantable sensors, vital sign monitors, infusion pumps, and imaging equipment. These devices generate vast amounts of data related to patient health metrics, such as heart rate, blood pressure, glucose levels, and electrocardiogram (ECG) readings. OT systems within medical devices capture and process patient data at the point of care. This data is then transmitted to IT infrastructure through wired or wireless networks, where it is stored, aggregated, and analyzed. OT devices often rely on standard IT communication protocols such as Wi-Fi, Bluetooth, or cellular connectivity to transmit data securely to backend IT systems located in hospitals, clinics, or cloud-based platforms. Connected healthcare systems leverage cloud computing and data analytics to process and analyze large volumes of patient data efficiently. Cloud-based platforms provide scalable infrastructure for storing and processing healthcare data, enabling healthcare providers to access critical information anytime, anywhere [3]. Data analytics tools and algorithms applied to IT systems help healthcare organizations derive insights from patient data, such as identifying trends, predicting disease outbreaks, and optimizing treatment protocols.

Connected healthcare systems raise cybersecurity and privacy concerns related to the protection of patient data and the security of medical devices. IT and OT professionals collaborate to implement robust cybersecurity measures, including encryption, access controls, network segmentation, and intrusion detection systems. OT security measures focus on securing medical devices against vulnerabilities and unauthorized access, while IT security measures safeguard patient data stored in electronic health records and transmitted over networks [1, 3]. This has led to the exploration and adoption of innovative technologies, such as Artificial intelligence (AI), to enhance the effectiveness and efficiency of cybersecurity measures. AI refers to the ability of machines to simulate human intelligence and perform tasks that typically require human intelligence, such as learning, reasoning, problem-solving, and decision-making. In the context of cybersecurity, AI technologies have shown immense potential in bolstering defences, identifying vulnerabilities, detecting anomalies, and responding to cyber incidents in real-time [1, 2]. Machine learning (ML), deep learning (DL), natural language processing (NLP), and data analytics are just a few of the methods and applications included in the field of AI in cybersecurity. These technologies enable systems to autonomously analyse vast amounts of data, identify patterns, and make informed decisions, augmenting the capabilities of cybersecurity professionals and improving overall security posture [4].

1.2 Significance of AI in Cybersecurity

Cybersecurity holds immense significance in the digital era, and its importance cannot be emphasized enough. A single security breach in our interconnected society can have far-reaching consequences. The brilliance of artificial intelligence lies in its ability to eliminate the need for explicit programming before performing a task. However, it is still feasible to develop a machine equipped with pre-programmed algorithms that utilize its intelligence to accomplish specific tasks. By harnessing the power of AI, we can detect and identify cyber-attacks originating from various web platforms and even reputable websites with stringent security measures in place [1, 4]. The real-time response and automation capabilities of AI enable swift action against security incidents, reducing the window of vulnerability. AI's significance in cybersecurity lies in its capacity to provide intelligent, automated, and adaptive solutions, fortifying digital environments against an ever-changing threat landscape and protecting them against any threats as shown in Fig. 1.

Fig. 1 A padlock symbolizing cybersecurity

2 Hack Alert: Navigating Cyber Threats

In an increasingly interconnected digital landscape, the prevalence of cybersecurity threats has become a critical concern for individuals and organisations alike. As we navigate the vast expanses of the online world, we encounter not only unprecedented opportunities but also an array of sophisticated threats poised to exploit vulnerabilities.

2.1 Cybersecurity Threats

There have been many kinds of cyberthreats throughout the past ten years. Cybersecurity threats pose constant challenges to individuals, organizations, and nations in our interconnected digital age. These threats encompass a wide range of malicious activities aimed at exploiting vulnerabilities and causing harm to systems, networks, and data. The top cyberthreats we face today, include:

a. Denial of Service (DoS) attacks aim to saturate a target system's processing power by flooding it with many requests that must be processed rapidly. These attacks can be executed in several ways: a lone attacker may initiate a distributed-style DoS (DDoS) attack by flooding the victim system with fabricated internet traffic packets, navigating around security barriers along the route [1]; a DDoS attack might involve numerous attacker machines, which increases the effect it has on the victim computer. As attacker tools become easily accessible, DoS attacks are growing in complexity and evading detection is becoming more challenging [5, 6].
b. Man-in-The-Middle (MiTM) attacks are traditional cyberattacks that involve intercepting data being sent across a communication channel between two authorised parties. The attacker establishes a physical or virtual barrier between the two parties so that they cannot communicate with one another. The attacker then pretends to be party A to speak with party B by intercepting their communication and replacing it with messages that are malicious or tampered with [6]. IP address spoofing is one of the attack's variant implementations, in which a hostile actor deceives lawful systems into believing it is a trustworthy entity to get access to the system [1, 5].
c. Phishing: Those involve creating emails that appear respectable and sending them to reliable systems that allow you to trick unsuspecting end users into clicking a link and disclosing non-public records. Such attacks use social engineering strategies to trick clients into believing emails are actual and gaining their consideration [5, 6].
d. Spear phishing is described as a properly deliberate attack that accommodates a radical historical past test on potential sufferers through the malicious actor, followed with the aid of the creation of emails that appear extraordinarily valid

and regularly contain trustworthy electronic mail addresses within the "from" box [1].

e. Drive-by attacks: These are carried out by hostile actors that browse the web looking for websites that are vulnerable so they can insert harmful scripts into the web servers. Visitors to the website eventually become infected with the malware, which causes system compromise, the release of sensitive data, and other harm [1, 5, 6].
f. Password attacks: This can be done by monitoring keyboard activity, brute force hacking, or using widely shared passwords and creation of advanced passwords using AI technology, etc. [5].
g. Birthday attacks: To ensure data integrity, a message hash, also known as a message digest, can be generated using well-known techniques such as Secure Hash Algorithm-1 (SHA-1). This technique produces a fixed-length hash value regardless of the original message's length. However, there is a security vulnerability known as the birthday attack, where an attacker attempts to find two different messages that result in the same hash value. This can lead to the substitution of the original message with another one that produces an identical hash value [5, 6]. Therefore, data loss and disruptions in systems and services can occur. In such attacks, artificial intelligence is employed to identify random messages that yield the same hash value as a legitimate message. This poses a significant threat to the integrity of data and the smooth functioning of systems [1].
h. Malware attack: One of the major challenges faced by web hosting companies is the potential for viruses to spread through their hosted websites. According to Symantec's 2016 Threat Report, around 78% of websites contain significant vulnerabilities that can be exploited by attackers to execute malicious code without any user involvement [5]. To bolster the defences of these websites, implementing the right security controls such as web proxies, firewalls, and intrusion detection systems becomes crucial. However, there exists a delicate balance between ensuring the appropriate level of security protections and maintaining the usability of the hosted websites. The more user-friendly and accessible a website is, the wider its vulnerability surface becomes. This creates a significant dilemma in finding the right equilibrium between security and usability to safeguard against potential attacks [1, 6].

2.2 *Traditional Cybersecurity Techniques*

Traditional cybersecurity techniques form the backbone of any robust cybersecurity strategy. These foundational methods include firewalls to control network traffic, antivirus software to detect and remove malware, intrusion detection and prevention systems for threat monitoring, access controls to manage user permissions, encryption for data protection, regular patching to address vulnerabilities, network segmentation for containment, security policies and training for user awareness, backup and disaster recovery for resilience, and security audits to identify weaknesses [2]. While these techniques provide a fundamental defence against cyber threats, the dynamic nature of cybersecurity requires ongoing adaptation to emerging risks, often involving the integration of advanced technologies and practices. Here are some key traditional cybersecurity techniques:

a. Game theory: Cybersecurity has previously utilised this. The victim's computer as the other participant in the game, with the malicious actor as the first player. Every participant strives to optimise their own motivation by strategically making moves that they justify as rational steps towards achieving their goal. The actions of each participant can either be predicted in advance or remain a secret. An illustration of a game would be a smart grid setting where the defender tries to maintain connectivity between various entities while the attacker tries to interfere with communication between a power system and a residence. Both the attacker and the defender would use techniques to accomplish their respective aims at each stage of the game [1].
b. Rate control: DoS and DDoS attacks target the availability of systems. By lowering the amount of incoming network traffic, performing basic traffic throttling, and revising permission lists, rate-control approaches can reduce the negative effects on the operation of such systems when they are under assault [1].
c. Anomaly based intrusion detection: By using this method, an illustration that illustrates what might be considered the norm is produced. The models may be based on rules, governmental regulations, mathematical models, and statistical methods. Changes from the ordinary are viewed as assaults. Such a strategy has the advantage of not relying on signature patterns and reduces the administrative burden of signature collection compared to signature-based detection [1].
d. Autonomous systems: As in the bio autonomic nervous system (BANS), these can self- heal, self-protect, and ensure dependability. The four different modules that make up this system are called Cyber Neuron, Cyber Axon, Peripheral Nerve, and Central Nerve. Malware and spyware are protected from Cyber Neuron. A sophisticated tool to repair spyware and virus damage is called Cyber Axon. Like how Peripheral Nerve creates a communication link between numerous cyber neurons installed on various devices, it offers a strong defence against DoS/DDoS attacks. Finally, Central Nerve acts as a knowledge store for potential threats and for communicating information to other security tools. It is suggested that peripheral nerves work together to defend the body against DoS and DDoS attacks through cooperation between devices within the network [1].
e. End user security controls: In-built security is preferred over add-ons for today's end-user devices, including cell phones, iPads, and smart portable devices. While some vendors try to promote automatic updates, Sometimes, end users neglect to update their devices with the latest security patches, leaving their devices vulnerable to potential threats. The WannaCry ransomware attack serves as a prime example of the consequences that arise when vendor-provided security fixes are not installed on all user devices. In many cases, people are unaware of

the potential risks associated with not applying these fixes. While some users may possess the knowledge, there are instances where they either fail to take the necessary actions to secure their devices or follow incorrect procedures, inadvertently exposing their devices to additional vulnerabilities. To address this issue, the concept of "out of sight" security has been proposed, where providers automatically deliver updates to end-user devices without requiring their explicit involvement. However, the challenge lies in ensuring that these security updates effectively safeguard against new and emerging attacks, including zero-day attacks, while seamlessly integrating with all existing software on the user's device. Software developers would need to carefully design and implement security updates to address these concerns and maintain the overall integrity and functionality of the end-user's device [1].

2.3 Innovative Approaches to Cybersecurity

The study of AI in computer science aims to imitate and increase human intelligence in machines through the development of theories, methodologies, techniques, and systems. AI seeks to imbue robots with human intelligence. AI may be implemented via machine learning, which uses algorithms to examine and learn from data. The technique of machine learning employs deep learning technology, which enables the development of AI. The foundation of AI is the correct description of human intellect, which enables computers and/or software to replicate it [2, 7].

Cyberattacks on Cyber-physical systems are growing as AI technologies get extra state-of-the-art and pervasive, taking advantage of the hyperlink among bodily and cyber elements. A couple of actors are concerned in the hazard surroundings, and attackers search for numerous vulnerabilities with which to launch their attacks. These attacks range from the sophistication and complexity of advanced continual threats to the use of our on-line world for malevolent functions and the commercialization of cybercrime.

2.3.1 Machine Learning

Machine learning (ML) is a system used to apply AI algorithms to dissect data, learn from the data, and make opinions about real-world events. ML systems can be divided into (i) systems for original training on the dataset; (ii) systems formerly trained for after decision- timber. Given the large quantum of data available, there is a strong demand for the operation of ML ways uses a variety of methods to address data issues. Considering the factors involved in the learning process; the kind of algorithm relies on the issue to be addressed. In the age of digital transformation, machine learning is an important subject of study for AI-based cyber security [2, 7].

Machine learning techniques are often categorised into two main types: supervised learning and unsupervised learning. In supervised learning, data samples are first classified into predefined classes, such as harmful or genuine, and then labelled accordingly. This process typically involves manual effort, where human experts recognize patterns in the data and assign appropriate labels [7]. The labelled data is then used to train an algorithm, which builds a mathematical model capable of classifying new data samples based on the learned patterns. On the other hand, unsupervised learning does not require data labelling or training. Instead, algorithms analyse the coherence or dispersion among data samples, systematically generating classes and categorising them based on the degree of coherence within each class and the modularity across classes [1, 8].

Machine learning plays a crucial role in detecting and preventing targeted attacks on industrial control systems. By training machine learning models to identify unusual activities that align with targeted attacks, it becomes possible to proactively recognize and prevent such attacks from occurring automatically. Integrating machine learning into anomaly detection systems can enhance the capabilities of these technologies. With machine learning, you can identify anomalous behaviour based on data patterns, or develop your own machine learning models that learn from your data [6]. In the context of network traffic, machine learning helps detect and evaluate abnormal behaviour, empowering network-based anomaly detection systems to respond appropriately like filtering out problematic traffic [7, 8].

Machine learning techniques as shown in Fig. 2 each play distinct roles in addressing various challenges within the field of cybersecurity. Supervised learning is employed in tasks where the algorithm is trained on a labelled dataset, learning to map input data to predefined output labels. and in developing models for malware detection by learning from labelled datasets containing examples of both benign and malicious software [8]. Unsupervised learning is utilized when the algorithm is given unlabelled data and must identify patterns or groupings on its own. Reinforcement learning involves an agent learning to make decisions by interacting with an environment and receiving feedback in the form of rewards or penalties and to respond to cyber threats in real-time by adapting to changing attack patterns. Semi-supervised learning combines both labelled and unlabelled data for training, allowing the model to learn from a limited set of labelled examples and a larger set of unlabelled data [6, 9].

In the context of anomaly detection, the model can be trained using supervised learning techniques on a dataset that includes both normal (non-anomalous) and anomalous instances. Anomaly detection relies on four crucial components: input, training data, model parameters, and output. The input refers to the sequence of observations where we aim to predict an abnormal event. These observations encompass various characteristics like TCP port numbers, HTTP header fields, and IP addresses found at a company's edge routers. The training data consists of annotated sequences of observations, often containing instances of anomalous events. Model parameters play a vital role as they define the training algorithm. They encompass normalisation parameters, anomaly detection sensitivity, and the detection threshold. These parameters determine how effectively the anomaly detector can identify an event and whether it exhibits overconfidence or under- confidence. Moreover, model parameters govern the behaviour of the anomaly detector when it detects false positives.

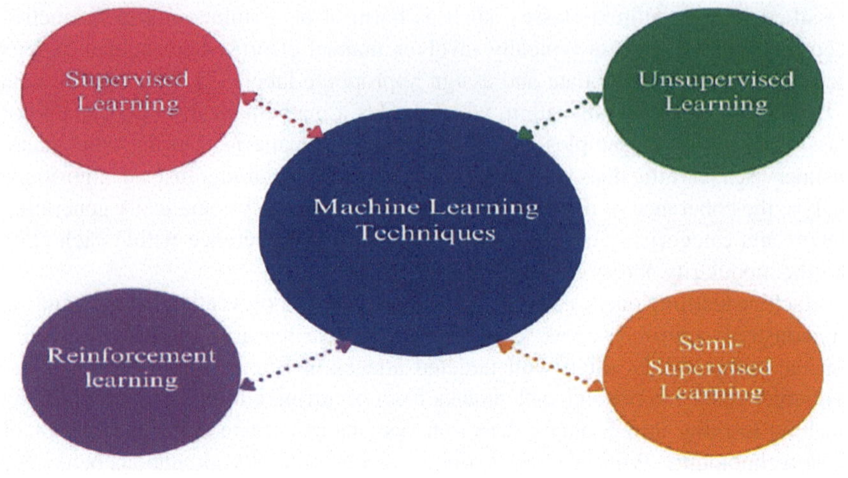

Fig. 2 Machine learning techniques

The outputs of an anomaly detector include confidence and hypotheses. Confidence signifies the likelihood of an anomalous event occurring within a given sequence. Hypotheses, on the other hand, provide potential explanations for the event, enabling the anomaly detector to identify patterns or sets of patterns in the data [7, 8].

2.3.2 Deep Learning

Other approaches have been suggested to improve detection accuracy. A more complicated subset of ML is known as deep learning. Researchers describe DL as a neural network with at least three hidden layers as shown in Fig. 3.

Deep learning (DL) emulates the functioning of the human brain when it comes to processing data and identifying patterns for decision-making purposes. It mimics the operations of neurons and intelligent thinking by processing signals [9]. The performance of neural networks is continuously improved by constructing larger and more complex networks and training them with vast amounts of data.

In the realm of cybersecurity, the generation of an enormous volume of data daily presents an opportunity for DL to be misused. DL algorithms excel in handling large datasets, which sets them apart from traditional machine learning (ML) approaches. DL models also support distinct types of learning, including supervised learning, unsupervised learning, and reinforcement learning, like ML techniques. Commonly employed DL algorithms in the field of cybersecurity include ensembles of feedforward neural networks, convolutional neural networks, recurrent neural networks, deep belief networks, stacked autoencoders, generative adversarial networks, constrained Boltzmann machines, and DL networks [9].

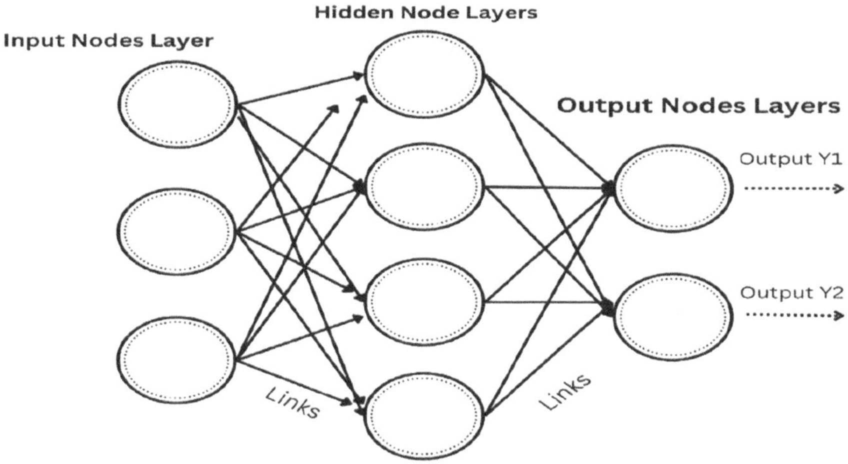

Fig. 3 Deep neural network architecture

Deep learning is revolutionizing cybersecurity by harnessing the power of advanced algorithms to bolster defence mechanisms. Its prowess lies in the thorough analysis of extensive datasets, enabling the detection of intricate patterns indicative of cyber threats. Particularly adept at recognizing both known and unknown threats, deep learning provides a proactive defence through anomaly detection and behavioural analysis. In the realm of malware and phishing, it excels by identifying evolving forms of malicious software and detecting subtle indicators of phishing attempts. Furthermore, its automated response capabilities ensure swift action, mitigating potential damage once a threat is identified. Deep learning's contribution extends to sophisticated behavioural analysis, enabling the detection of unusual activities within networks or systems. Deep learning is a cornerstone in modern cybersecurity, offering a robust and adaptive approach to fortify digital environments against the ever-evolving landscape of cyber threats [9].

3 Applications of AI in Cybersecurity

Artificial Intelligence (AI) plays a pivotal role in revolutionizing cybersecurity across various domains. It excels in threat detection by analysing patterns and anomalies in user behaviour and network activities, offering a proactive defence against evolving cyber threats. AI enhances endpoint security, providing real-time protection against malware and ransomware. Its behavioural analysis capabilities aid in identifying insider threats and sophisticated attacks that may elude traditional security measures.

3.1 Threat Detection

AI methods play a crucial role in identifying potential cybersecurity threats and preventing attacks before they can cause harm. It does this by analysing an extensive dataset of cybersecurity events and identifying patterns indicative of malicious activity. The model used in this process builds on previous data monitoring and indications of compromise (IOCs) that help detect, identify, and respond to threats in real time. When similar patterns are detected, the model automatically recognizes and labels them [7]. Machine learning algorithms use IOC datasets to identify and classify diverse types of malwares. Behavioural analysis, on the other hand, uses clustering and bracketing algorithms to analyse large numbers of malware. These patterns can also be used to automate the process of detecting and classifying new threats. Security teams and automated systems benefit from AI-based threat detection as they can respond quickly and prevent potential threats from turning into serious security incidents. These systems can evaluate data from multiple sources simultaneously, allowing them to detect and track threats across disparate systems and networks [10].

Depending on the data and algorithms used, AI-based threat detection systems can identify a wide range of threats, such as malware, phishing schemes, and other online dangers. In the realm of physical security, AI can even identify suspicious activities or behaviours in video surveillance footage, such as unauthorised entry or theft. However, there are challenges in anomaly detection, as anomalies may be rare and difficult to capture in training data, resulting in imbalanced datasets. To address this, feature engineering plays a crucial role in extracting relevant features that capture the characteristics of normal behaviour and anomalies. Furthermore, ongoing monitoring and adaptation of the models are necessary to ensure the effectiveness of the anomaly detection system.

Darktrace is a leading cybersecurity company that specializes in AI-powered threat detection and response. The company's flagship product, the Darktrace Immune System, is an AI-driven platform designed to autonomously detect and respond to cyber threats in real-time. Darktrace's AI algorithms are inspired by the human immune system. The platform continuously learns the 'pattern of life' for every user and device within an organization's network. This includes understanding typical behaviors, communication patterns, and network traffic. Darktrace provides coverage across all types of environments, including on-premises networks, cloud infrastructure, IoT devices, and email systems [8, 9]. This comprehensive coverage allows organizations to protect their entire digital estate from cyber threats. Darktrace's AI models continuously learn and adapt to new threats and changes in network behavior. This adaptability is crucial in the ever-evolving landscape of cybersecurity, where attackers constantly develop new techniques to bypass defenses.

3.2 Vulnerability Management

AI can assist in the analysis of discovered vulnerabilities to determine their severity and potential impact. By leveraging machine learning algorithms, AI systems can analyse vulnerability data, historical attack patterns, and contextual information to assess the likelihood of exploitation and the potential consequences of a successful attack. This helps prioritise vulnerabilities for remediation based on their criticality, enabling organisations to allocate resources effectively. AI can aid in predicting the impact of applying patches to vulnerabilities [11].

Machine studying algorithms can examine ancient statistics on patch deployments, machine configurations, and the presence of comparable vulnerabilities to predict the likelihood of successful patching and the potential side effects or compatibility issues that may arise. This helps organisations make informed decisions regarding patch deployment strategies, minimising the risk of disruptions or introducing new vulnerabilities. AI can be used to model and simulate potential threats and attack scenarios based on known vulnerabilities. By leveraging historical attack data and machine learning algorithms, AI systems can generate realistic threat scenarios, identifying potential attack vectors and assessing the impact of successful exploitation. This allows organisations to proactively address vulnerabilities and implement countermeasures to mitigate potential risks [11].

AI-based vulnerability assessment also comes with challenges. The accuracy of AI models depends on the quality and diversity of the training data. Lack of comprehensive and up-to-date vulnerability data can impact the performance of AI algorithms. Additionally, adversarial attacks aimed at manipulating vulnerability assessment systems pose a concern. Attackers may attempt to obfuscate or hide vulnerabilities to evade detection by AI-based tools. Ongoing research and development are necessary to address these challenges and ensure the reliability and effectiveness of AI-driven vulnerability assessment. Let's consider a scenario where a critical vulnerability in a widely used software product was announced during Patch Tuesday. An AI-empowered ASM system continually monitors the web for threats related to this vulnerability. It detects a threat actor on an underground forum discussing the exploitation of this vulnerability.

The AI system immediately alerts the system admin with a high-priority notification, providing the following information:

- Details of the specific vulnerability.
- The threat actor's username and their previous activity.
- An estimation of the threat actor's capabilities and motivations.
- Recommendations for immediate patching and isolating affected systems.

This real threat displayed using AI in attack surface management allows the system admin to take prompt action to protect the organization's assets. The AI-driven visualization of the attack surface management dashboard highlights the criticality of the situation and provides a clear understanding of the attack surface, ultimately contributing to a more secure and efficient response [11].

4 AI in Cybersecurity: Benefits

Artificial Intelligence (AI) significantly enhances cybersecurity by providing advanced capabilities in threat detection, incident response, and overall defence mechanisms. AI-powered tools analyse vast amounts of data in real-time, detect anomalies, and autonomously respond to cybersecurity incidents. Machine learning models enable predictive analysis, phishing detection, and adaptive security measures, while also reducing false positives. AI's role extends to user authentication, access control, and the augmentation of cybersecurity skills, contributing to a more robust and proactive cybersecurity posture [1, 12].

4.1 Enhanced User Authentication

a. Biometric Authentication: AI-driven biometric authentication systems make use of advanced technologies like fingerprint recognition, facial recognition, voice recognition, and iris scanning. These innovative systems enable us to uniquely identify individuals based on their distinctive biological characteristics. By employing artificial intelligence, these authentication systems analyse and interpret biometric data to verify a person's identity in a highly accurate and efficient manner. Whether it is scanning fingerprints, analysing facial features, recognizing voices, or examining iris patterns, AI-powered biometric authentication systems provide a secure and reliable means of confirming who we are, ensuring that our personal information and access to sensitive data remain protected [12]. These systems can accurately verify a user's unique physiological characteristics to grant access. AI algorithms analyse biometric data and match it against pre-registered templates, providing a highly secure and convenient authentication method. AI algorithms can analyse and recognize unique behavioural patterns of users, such as typing speed, mouse movements, gesture patterns, and navigation behaviour. By continuously monitoring and analysing these behavioural biometrics, AI systems can authenticate users based on their individual patterns, adding an extra layer of security over traditional authentication methods [1].
b. Continuous Authentication: AI enables continuous authentication by continuously monitoring user behaviour and analysing real-time data. This approach reduces the reliance on static, one-time authentication and adds an extra layer of security. If AI detects suspicious or anomalous behaviour, it can prompt for additional authentication or trigger an alert for potential unauthorised access.AI algorithms can analyse contextual information, such as location, device information, network behaviour, and access patterns, to determine the legitimacy of user authentication requests. By comparing the current authentication attempt with historical data and user profiles, AI can assess the likelihood of fraudulent activity and apply appropriate authentication measures [1, 12].

c. Risk-based Authentication: AI can assess the risk level associated with each authentication attempt by considering numerous factors, including user behaviour, device security posture, location, and network information. Based on the risk level, AI systems can dynamically adapt authentication requirements, prompting for additional verification steps or applying stronger authentication methods when the risk is higher [1]. When a user attempts to authenticate or access a system, the AI system assesses the risk associated with the request based on deviations from the user's normal behavior and other contextual information. For example, if a user typically logs in from a specific location during business hours but suddenly attempts to log in from a different country at an unusual time, it might raise a higher risk score. Based on the risk assessment, the system dynamically adjusts the authentication requirements. For low-risk requests, the user may be granted access with minimal friction, such as a single-factor authentication. However, for high-risk requests, the system may require additional verification steps, such as multi-factor authentication or biometric authentication. One real-world example of RBA utilizing AI in cybersecurity is the use of behavioral biometrics by financial institutions and online service providers. These organizations deploy AI-powered systems that analyze user behavior, such as typing patterns, mouse movements, and touchscreen interactions, to create unique behavioral profiles for each user. When a user attempts to log in, the system compares their current behavior to their established profile and assigns a risk score based on any deviations or anomalies detected [12].

4.2 Adaptive and Self Learning Capabilities

Based on feedback loops and new data, AI algorithms continuously learn and advance. AI systems may continuously improve their models, rules, and algorithms by keeping track of the results of security measures. This recurrent development ensures continuing optimisation of cybersecurity measures by enabling AI to alter and respond to changes in the threat landscape. Because AI is self-learning, it is capable of predictive analytics, which predicts threats and weaknesses based on past data and patterns. AI can direct proactive security measures and enable firms to anticipate new risks by spotting developing trends and forecasting future attack vectors [13, 14].

Analysing a variety of variables, such as user behaviour, network circumstances, device attributes, and contextual information, AI systems may continuously assess danger levels. AI can adjust security measures, such as authentication standards or access limits, to meet the perceived degree of danger by dynamically analysing risk. Organisations can effectively allocate resources and rank security measures by using this adaptive risk assessment. Incident response and remediation processes can be automated. AI systems can autonomously respond to and mitigate common risks by studying historical event data and learning from previous responses, cutting response times, and lessening the effect of security incidents. AI can also learn from effective replies to better incident management as a future event handling.

4.3 Real Time Threat Intelligence

AI algorithms can utilise a diverse range of sources to collect and consolidate data, such as threat feeds, security vendors, open-source intelligence, monitoring the dark web, honeypots, and security sensors. These algorithms are designed to intelligently gather information from these various sources, bringing together a comprehensive dataset for analysis and decision-making in the realm of cybersecurity. By leveraging the power of AI, organisations can effectively collect and collate data from multiple channels, enhancing their ability to detect and respond to potential threats more efficiently [15]. As a result, threat intelligence is continuously updated and complete thanks to these algorithms' ability to automatically collect and analyse data in real-time. AI makes it possible to automate data analysis, which enables the processing of huge volumes of threat intelligence data in real-time. AI algorithms can quickly analyse and extract pertinent data, including malware signatures, behavioural patterns linked to known threats, and indicators of compromise (IOCs) [16].

Organisations can receive early warnings about new risks and attack campaigns thanks to real-time threat intelligence. Organisations may proactively identify and respond to threats in their early phases, before they do major damage, by utilising AI's capabilities in pattern recognition and anomaly detection. This proactively defensive stance aids in risk reduction and lessens the effects of possible attacks. AI algorithms can analyse threat data in the context of an organisation's specific environment, including network infrastructure, applications, and user behaviour. By combining internal data with external threat intelligence, AI can provide highly relevant threat intelligence customised for an organisation's unique security environment [15, 16].

4.4 Anomaly Detection

AI can detect various types of anomalies or irregularities within a network or system, including:

a. Unusual User Behaviour: AI algorithms can analyze login timestamps and geographic locations associated with user logins. Deviations from typical login times or unexpected login locations may indicate unauthorized access attempts or compromised accounts. AI-driven security solutions can detect instances where users attempt to escalate their privileges beyond their authorized levels. Unauthorized privilege escalations could indicate malicious intent or compromised user accounts [17]. They also can monitor user interactions with sensitive data and identify deviations from normal data access patterns. Unusual data access behaviors, such as accessing files or databases that users typically do not interact with, may indicate data theft or unauthorized access attempts.AI algorithms can analyse user account behaviors, such as login frequency, session duration, or access patterns, to identify suspicious activities indicative of compromised accounts or insider threats. Unusual account behaviors may include simultaneous logins from multiple locations, irregular access patterns, or unusual account activity during off- hours.

b. Unauthorized Access Attempts: AI-powered intrusion detection systems (IDS) play a crucial role in monitoring network traffic and analyzing access logs to identify unauthorized access attempts [10]. These systems employ sophisticated algorithms to detect patterns indicative of suspicious login activities, brute-force attacks, or attempts to gain unauthorized access to sensitive resources. By correlating multiple indicators of compromise (IOCs), AI algorithms can accurately identify and alert security teams to potential unauthorized access attempts, enabling timely investigation and response.
c. Malware Activity: AI-based malware detection systems leverage machine learning algorithms to analyze file attributes, behavior patterns, and network communications to detect malware infections. These systems can identify anomalies such as unusual file behavior, suspicious process executions, or aberrant network traffic, which may signal the presence of malware such as ransomware, Trojans, or botnets [10]. By continuously monitoring for these indicators, AI-powered malware detection systems can proactively detect and mitigate malware threats, helping organizations protect their systems and data from malicious attacks.
d. Network Anomalies: AI systems can analyze the behavior of users and devices within a network to identify anomalies. For example, sudden changes in the volume or type of data being transmitted, unusual login patterns, or unexpected communication between devices can all be signs of potential security threats.AI and machine learning algorithms are increasingly being used to detect anomalies in network traffic that might signify a security breach or intrusion. These algorithms can learn the normal patterns of network behavior and raise alerts when they detect deviations from these patterns [10, 17]. AI excels at recognizing complex patterns in large datasets, which is essential for detecting anomalies in network traffic. By continuously monitoring and analyzing network data, AI systems can quickly identify deviations from normal behavior, such as a sudden spike in traffic or unusual communication patterns.

5 Challenges and Limitations of AI in Cybersecurity

While AI has the potential to assist in cybersecurity, there are certain limitations that prevent it from becoming a widely adopted tool in this field. One of the key challenges we encounter when integrating AI into cybersecurity practices is the considerable demand for resources. Training AI models for cybersecurity applications necessitates substantial computational resources. These models often require extensive datasets for training, which means we must deploy powerful computing infrastructure capable of handling large-scale data processing and complex algorithms.

Furthermore, the deployment of AI-powered cybersecurity solutions often requires infrastructure investments. Integrating AI models into existing security frameworks may require upgrades or modifications to network architectures, data storage systems, and endpoint devices. Additionally, organizations must ensure robust cybersecurity measures are in place to safeguard AI training data, model parameters, and inference results from unauthorized access or tampering [1, 16].

Additionally, cybersecurity companies need access to diverse datasets containing malware code, benign code, and anomalies to train AI devices effectively. Acquiring these accurate records can be time-consuming and costly, which may be prohibitive for some companies. Cybersecurity datasets are often characterized by noise, incompleteness, and inconsistencies, stemming from factors such as data collection methods, sensor inaccuracies, and human error [18]. This inherent data variability poses challenges for AI algorithms, as they may struggle to distinguish meaningful patterns from irrelevant or misleading information. Ensuring the quality and integrity of cybersecurity data through heavy data pre-processing, cleansing, and validation processes is essential to mitigate the risk of introducing biases or errors into AI-driven security systems [1, 19]. Another important limitation of AI in cybersecurity is its vulnerability to manipulation and misuse. AI systems are programmed computer code that can be controlled and potentially weaponized if manipulated by individuals with malicious intent. Even a minor alteration in the code can transform a well-intentioned AI system into a dangerous weapon. AI algorithms can be manipulated or deceived by malicious actors seeking to evade detection or exploit weaknesses in our security defenses. Adversarial attacks, where attackers craft input data to intentionally mislead AI systems, pose a formidable threat to the integrity and reliability of AI-driven security solutions. By exploiting vulnerabilities in AI algorithms or leveraging adversarial techniques such as data poisoning or model inversion, attackers can subvert the effectiveness of AI-based threat detection mechanisms, leading to false positives, false negatives, or evasion of detection. The proliferation of AI-generated synthetic content, such as deepfake videos, text, and audio, poses challenges for cybersecurity professionals tasked with detecting and mitigating disinformation campaigns, impersonation attacks, and other forms of AI-mediated deception. The ability to generate realistic- looking fake content using AI algorithms complicates efforts to discern between genuine and manipulated information, raising concerns about trust, authenticity, and accountability in digital environments [19]. Therefore, the potential for AI technology to be subverted and used for nefarious purposes is a significant cybersecurity concern. While AI systems can be trained to detect cyber threats and malicious malware, they are not capable of replacing humans entirely. AI is designed to perform specific tasks and lacks the comprehensive understanding and adaptability of human experts. Some threats, especially those that closely resemble genuine messages, may be difficult for AI to distinguish accurately. Furthermore, the ever-evolving nature of cyber threats presents a challenge for AI systems, as cybercriminals constantly improve their techniques and can create sophisticated threats that AI struggles to detect [19, 20].

Thus, human creativity and expertise are still crucial in combating cyber threats. The complexity of AI technology is another limiting factor. Not everyone in society is technologically savvy or possesses a deep understanding of various AI models. Implementing and utilising this technology effectively can be challenging, resulting in underutilization, and leaving vulnerabilities that criminals can exploit. The intricate nature of AI technology also demands a significant amount of knowledge to operate it correctly, which many individuals lack. As a result, some businesses continue to be vulnerable as, for lack of expertise, their systems might not be properly optimal. Furthermore, a major barrier to the general acceptance of AI technology is its cost. Not every organization has access to the tools required to use AI technology to secure data. The cost of implementing and maintaining AI systems may prevent many organizations from adopting them. Due to the lack of users, it is difficult to evaluate the technology's full potential and there is a greater risk to businesses that cannot afford to implement it [19]. In conclusion, while AI offers promising potential in enhancing cybersecurity, it is essential to acknowledge its limitations. The resource-intensive nature of building and maintaining AI systems, the susceptibility to manipulation, the need for human expertise, the complexity of the technology, and the cost of implementation all contribute to the challenges of fully harnessing AI's capabilities in cybersecurity.

6 Future-Proofing Security: AI in Cyber Trends

The field of AI in cybersecurity is constantly evolving, and we can expect to see exciting developments and trends soon. AI will continue to play a crucial role in security analytics by using advanced machine learning and deep learning algorithms to analyse large volumes of security data. This will enable security teams to detect and respond to emerging threats more effectively, employing techniques like threat detection, anomaly detection, and behavioural analytics [4, 21].

One area where AI will make a significant impact is in enhancing threat hunting capabilities. By automating the detection of complex attack patterns and indicators of compromise, security analysts will be able to quickly sift through vast amounts of data and proactively hunt for potential attacks. This proactive approach can help prevent considerable damage caused by cyber threats. As privacy concerns become more prominent, there will be a focus on developing privacy- preserving AI techniques in cybersecurity. Technologies like federated learning and secure multi- party computation will allow organisations to collaborate and analyse data without compromising individual privacy. This will enable organisations to leverage the power of AI while ensuring the protection of sensitive information [22]. Another exciting development is the integration of AI with quantum computing. As quantum computing progresses, there will be exploration of how AI can leverage quantum algorithms to enhance cybersecurity. This integration may improve task efficiency such as cryptography and optimization, offering new possibilities for secure communication and data protection.

Additionally, as AI algorithms become more sophisticated, there will be a greater emphasis on developing explainable AI models in cybersecurity. Explainable AI aims to provide transparent and interpretable results, allowing security analysts to understand how AI-driven decisions are made. This transparency promotes better trust, accountability, and decision-making in security operations [21, 22]. Overall, the future of AI in cybersecurity holds promise, with advancements in threat detection, privacy preservation, quantum integration, and explain ability. These developments will empower security teams to stay ahead of evolving cyber threats and protect sensitive information more effectively.

7 Conclusion and Future Scope of AI in Cybersecurity

Artificial intelligence (AI) has emerged as a powerful tool in the field of cybersecurity, revolutionising how organisations detect, prevent, and respond to cyber threats. AI brings numerous advantages, including improved threat detection and response time, scalability and automation, adaptive and self-learning capabilities, real-time threat intelligence, advanced behavioural analytics, and enhanced user authentication. These advantages enable organisations to strengthen their security posture, proactively identify and mitigate threats, and improve overall cybersecurity effectiveness. AI-based approaches, such as machine learning and deep learning, have proven effective in various cybersecurity domains, including threat detection, vulnerability assessment, incident response, and user authentication. By leveraging AI technologies, organisations can analyse large volumes of data, detect patterns and anomalies, identify vulnerabilities, and make informed decisions to protect their systems and data [23, 24].

While AI offers significant benefits, challenges exist. Issues such as data quality, adversarial attacks, explain ability, and workforce skills require attention and ongoing research. Addressing these challenges will be crucial to ensure the responsible and effective implementation of AI in cybersecurity. Looking ahead, future developments and trends in AI-driven cybersecurity include AI-enabled threat hunting, privacy-preserving AI, integration with quantum computing, explainable AI, AI-based deception technologies, autonomous security operations centres (SOCs), and enhanced user and entity behaviour analytics (UEBA). These advancements will further empower organisations to stay ahead of emerging threats, enhance security operations, and protect against evolving cyber risks [24, 25]. In conclusion, AI's role in cybersecurity is expanding, enabling organisations to bolster their defences, detect and respond to threats more effectively, and enhance overall cybersecurity resilience. By harnessing the power of AI, organisations can adapt to the ever-changing threat landscape and better safeguard their systems, data, and users in the digital age. However, it is important to strike a balance between leveraging the benefits of AI and addressing ethical considerations, privacy concerns, and the need for human oversight in cybersecurity operations.

References

1. Zeadally, S., Adi, E., Baig, Z., Khan, I.A.: Harnessing artificial intelligence capabilities to improve cybersecurity. IEEE Access **8**, 23817–23837 (2020). https://doi.org/10.1109/ACCESS.2020.2968045
2. Wiafe, I., Koranteng, F.N., Obeng, E.N., Assyne, N., Wiafe, A., Gulliver, S.R.: Artificial intelligence for cybersecurity: a systematic mapping of literature. IEEE Access **8**, 146598–146612 (2020). https://doi.org/10.1109/ACCESS.2020.3013145
3. Kumar, V., Singh, B., Sharma, S., Sharma, D.: A machine learning approach for predicting onset and progression-towards early detection of chronic diseases. J. Pharm. Neg. Res. (2022). https://doi.org/10.48047/nq.2022.20.22.NQ10187. Available: https://www.pnrjournal.com/index.php/home/article/view/10496/14691
4. Yampolskiy, R., Spellchecker, M.: Artificial Intelligence Safety and Cybersecurity: A Timeline of AI Failures (2016). https://doi.org/10.48550/arXiv.1610.07997
5. Li, J.: Cyber security meets artificial intelligence: a survey. Front. Inf. Technol. Electron. Eng. **19**, 1462–1474 (2018). https://doi.org/10.1631/FITEE.1800573
6. Thakur, K., Qiu, M., Gai, K., Ali, M.L.: An investigation on cyber security threats and security models. In: 2015 IEEE 2nd International Conference on Cyber Security and Cloud Computing, New York, NY, USA, pp. 307–311 (2015). https://doi.org/10.1109/CSCloud.2015.71
7. Razzaq, A., Hur, A., Ahmad, H.F., Masood, M.: Cyber security: threats, reasons, challenges, methodologies and state of the art solutions for industrial applications. In: 2013 IEEE Eleventh International Symposium on Autonomous Decentralised Systems (ISADS), Mexico City, Mexico, pp. 1–6 (2013). https://doi.org/10.1109/ISADS.2013.6513420
8. Shaukat, K., Luo, S., Varadharajan, V., Hameed, I.A., Xu, M.: A survey on machine learning techniques for cyber security in the last decade. IEEE Access **8**, 222310–222354 (2020). https://doi.org/10.1109/ACCESS.2020.3041951
9. Louridas, P., Ebert, C.: Machine learning. IEEE Softw. **33**(5), 110–115 (2016). https://doi.org/10.1109/MS.2016.114
10. Lauzon, F.Q.: An introduction to deep learning. In: 2012 11th International Conference on Information Science, Signal Processing, and their Applications (ISSPA), Montreal, QC, Canada, pp. 1438–1439 (2012). https://doi.org/10.1109/ISSPA.2012.6310529
11. Syed, R.: Cybersecurity vulnerability management: a conceptual ontology and cyber intelligence alert system. Inf. Manag. **57**(6), 103334 (2020). ISSN 0378-7206. https://doi.org/10.1016/j.im.2020.103334
12. Chan, L., et al.: Survey of AI in cybersecurity for information technology management. In: 2019 IEEE Technology & Engineering Management Conference (TEMSCON), Atlanta, GA, USA, pp. 1–8 (2019). https://doi.org/10.1109/TEMSCON.2019.8813605
13. Shah, S.W., Kanhere, S.S.: Recent trends in user authentication—a survey. IEEE Access **7**, 112505–112519 (2019). https://doi.org/10.1109/ACCESS.2019.2932400
14. Deng, L.: Artificial intelligence in the rising wave of deep learning: the historical path and future outlook [perspectives]. IEEE Signal Process. Mag. **35**(1), 177–180 (2018). https://doi.org/10.1109/MSP.2017.2762725
15. Yao, H., Mai, T., Xu, X., Zhang, P., Li, M., Liu, Y.: Network AI: an intelligent network architecture for self-learning control strategies in software defined networks. IEEE Internet Things J. **5**(6), 4319–4327 (2018). https://doi.org/10.1109/JIOT.2018.2859480
16. Lobato, A.G.P., Lopez, M.A., Sanz, I.J., Cardenas, A.A., Duarte, O.C.M.B., Pujolle, G.: An adaptive real-time architecture for zero-day threat detection. In: 2018 IEEE International

Conference on Communications (ICC), Kansas City, MO, USA, pp. 1–6 (2018). https://doi.org/10.1109/ICC.2018.8422622
17. Mishra, R., Sharma, D.: A PDF based scale-free topology construction model for wireless sensor networks. Wireless Pers. Commun. **131**, 2371–2397 (2023). https://doi.org/10.1007/s11277-023-10418-z
18. Wu, J., Haider, S.A., Soni, M., Kalra, A., Deb, N.: Blockchain based energy efficient multi-tasking optimistic scenario for mobile edge computing. PeerJ Comput. Sci. **8**, e1118 (2022). https://doi.org/10.7717/peerj-cs.1118.SCIE
19. More, R., Unakal, A., Kulkarni, V., Goudar, R.H.: Real time threat detection system in cloud using big data analytics. In: 2017 2nd IEEE International Conference on Recent Trends in Electronics, Information & Communication Technology (RTEICT), Bengaluru, India, pp. 1262–1264 (2017). https://doi.org/10.1109/RTEICT.2017.8256801
20. Chaudhary, H., Detroja, A., Prajapati, P., Shah, P.: A review of various challenges in cybersecurity using artificial intelligence. In: 2020 3rd International Conference on Intelligent Sustainable Systems (ICISS), Thoothukudi, India, pp. 829–836 (2020). https://doi.org/10.1109/ICISS49785.2020.9316003
21. Sedjelmaci, H., Guenab, F., Senouci, S.-M., Moustafa, H., Liu, J., Han, S.: Cyber security based on artificial intelligence for cyber-physical systems. IEEE Netw. **34**(3), 6–7 (2020). https://doi.org/10.1109/MNET.2020.9105926
22. Tao, F., Shoaib Akhtar, M., Jiayuan, Z.: The future of artificial intelligence in cybersecurity: a comprehensive survey. EAI Endorsed Trans. Creat. Tech. [Internet] (2021). https://doi.org/10.4108/eai.7-7-2021.170285
23. Morel, B.: Artificial intelligence and the future of cybersecurity. In: Proceedings of the 4th ACM Workshop on Security and Artificial Intelligence (AISec'11). Association for Computing Machinery, New York, NY, USA, pp. 93–98 (2011). https://doi.org/10.1145/2046684.2046699
24. Laato, S., Farooq, A., Tenhunen, H., Pitkamaki, T., Hakkala, A., Airola, A.: AI in cybersecurity education—a systematic literature review of studies on cybersecurity MOOCs. In: 2020 IEEE 20th International Conference on Advanced Learning Technologies (ICALT), Tartu, Estonia, pp. 6–10 (2020). https://doi.org/10.1109/ICALT49669.2020.00009
25. Capuano, N., et al.: Explainable artificial intelligence in cybersecurity: a survey. IEEE Access (2022). https://doi.org/10.1109/ACCESS.2022.3204171

Behavioural Decentralized Biometrics-Based Trust Score Authentication (BDB-TSA) Scheme to Assess Device Legitimacy in IoT

Ayasha and S. Athilakshmi

Abstract The exponential growth of the IoT has related billions of devices to the net and revolutionized an extensive range of worldwide corporations. The defence against cyberattacks, the protection of personal statistics, and the preservation of self-belief in IoT networks depend on shielding IoT devices. Traditional safety tactics are tough to execute on IoT gadgets because of their confined sources and widespread range of communique protocols. Since the life of a reliable device legitimacy assessment machine (LES) is essential to the continued dependability of IoT networks, it's far addressed. Due to variables, restricted sources, scalability challenges, numerous device sorts, and persistent development, authenticating the authenticity of related devices inside the Internet of Things (IoT) has its own precise set of challenges. Current processes are inadequate and can't offer a stable and efficient validation procedure. An authentication method that uses believe ratings generated from behavioural capabilities collectively with blockchain generation and minimally intrusive cryptographic tactics is supplied inside the current look as Behavioral Decentralized Biometrics-Based Trust Score Authentication (BDB-TSA). The cautioned technique offers an efficient, flexible, and secure way to check the legitimacy of IoT gadgets. It authenticates and verifies gadgets by integrating device-specific traits, historical statistics, and decentralized trust mechanisms. The capability to safely onboard gadgets and limit the right of entry to and transmit records all contribute to an environment of trust in IoT. This paper ran widespread simulations in several Internet of Things (IoT) environments to examine how the suggested method works nicely. The consequences show that in comparison to exclusive techniques, it is more effective, scalable, and attack-resistant. The simulation findings validate the precise validation approach's viability when applied to real Internet of Things (IoT) deployments.

Ayasha (✉) · S. Athilakshmi
Department of Computer Science, Christ College of Science and Management, Alambadi, Kolar, Karnataka, India
e-mail: ayasha@christcollegemalur.com

S. Athilakshmi
e-mail: athilakshmi.s@christcollegemalur.com

S. Mahajan et al. (eds.), *Smart Systems: Engineering and Managing Information for Future Success*, Information Systems Engineering and Management 22,
https://doi.org/10.1007/978-3-031-76152-2_7

Keywords Secure · Validation · Legitimacy · Internet of Things · Decentralized · Biometrics · Trust · Score · Authentication

1 Overview

A single, reliable validation method is essential to verify the authenticity of IoT machines in modern-day virtual surroundings. Due to the improved susceptibility of networked devices within the ever-increasing IoT surroundings to malicious intent, robust protection measures are critical [1]. This approach is vital to ensure the protection of IoT networks because of its novel technique for tool authentication [2]. The security of touchy information as it passes through the network of networked IoT gadgets, the mitigation of cyber threats and the prevention of unauthorized access are its most powerful factors [3]. By proscribing admission to the community to permitted gadgets only, the method establishes a layer of safety–critical to the IoT infrastructure's general dependability [4].

Furthermore, because the Internet of Things (IoT) is growing into critical sectors like healthcare, smart cities, and commercial enterprise automation, the reliability of tool validation is vital to protect from catastrophic effects [5]. In this manner, the Novel Secure Validation Scheme improves individual devices' safety and the overall stability and dependability of the developing Internet of Things environment [6]. In this era of growing virtualization, its implementation is an excessive first-class step in building a robust and constant infrastructure for the increasing network of interconnected devices [7]. A modern, steady validation scheme for evaluating the legitimacy of IoT devices offers a sparkling attitude at the pervasive safety problems impacting the IoT surroundings [8]. Concerns about interoperability and scalability are incredibly familiar when it comes to the implementation of this method [9]. The exponential growth of the Internet of Things (IoT) system makes the bold mission of supporting many devices with specific capabilities and communique protocols appear even more impossible [10]. The warned scheme may also have trouble integrating with exceptional varieties of IoT devices, possibly leading to compatibility issues and confined scheme utility in the IoT spectrum [11].

The virtual community typically develops reward systems and unique identification techniques to recognize members' contributions to the community. These mechanisms typically involve assigning points to members over time or using diamonds to indicate a member's social standing within the community. A symbolic suggestion is a more straightforward identification framework that facilitates trust delivery. Building a personal brand is always a sign that someone has gained recognition. Only in a social context can trust develop; the medium's social presence probably influences the social aspect of interpersonal communications. In addition to facilitating the development of other trust-building elements, social presence fosters trust on its own. The website can improve its social presence because of the online community's incentive system and unique identity, which can transmit more information and reduce ambiguity and equivocality in information exchanges. In the meantime, studies have indicated that the parties can establish more trust in one another, the more socially present they are. As a result, the VC incentive system is represented as the "reputation" dimensional element inside the collection of multidimensional elements that underpin the online trust paradigm.

Verifying the validity scheme's scalability is vital for ensuring gadgets' green and speedy affirmation in massive-scale IoT installations [12]. The mission is to strike a compromise between the practical constraints of actual-world global IoT configurations and the need for thorough safety features [13]. Achieving consistency among security, scalability, and interoperability is critical for the hit deployment of the Novel Secure Validation Scheme inside the continuously evolving Internet of Things surroundings [14]. Now, many cryptographic, biometric, and anomaly detection procedures are utilized by the Novel Secure Validation Scheme for Evaluating the Legitimacy of IoT Devices. Cryptographic methods, including digital signatures and certificates, may additionally prove the legitimacy of tool credentials and provide steady channels for verbal interaction [15]. Biometric authentication, which entails scanning a person's fingerprints or retina, is a better security degree for the IoT community. An additional layer of protection is supplied with anomaly detection algorithms, which screen tool conduct for developments and flag any unusual occurrences that could suggest safety worries [16].

The computational value of cryptographic methods is a primary trouble for IoT devices with limited access to sources. The accuracy of biometric authentication structures may be tormented by ambient elements, which similarly increases privacy issues while integrating them. Although effective, anomaly detection calls for complex algorithms and may be influenced by false positives and negatives; as a result, it requires continuous adjustment to prevent exceptional effects. Interoperability issues may hinder the validation method's integration throughout distinct IoT ecosystems, as personal devices may impose exclusive security requirements. Because of the exponential growth of IoT gadgets, scalability has become difficult, making green methods to confirm massive gadgets in actual time crucial. These worries must be resolved to efficiently implement a steady validation technique in the dynamic and challenging IoT system.

- To turn the tables on the safety troubles plaguing the IoT, this observation proposes a brand new technique of secure validation. Due to the diversity of IoT devices and the limitations of modern protection requirements, growing an honest technique to decide the legitimacy of connected devices is of extreme significance.
- Behavioural Decentralized Biometrics-Based Trust Score Authentication (BDB-TSA) is a modern-day authentication tool this study aims to offer. This approach relies on blockchain introduction, minimally intrusive cryptographic techniques, and rankings decided by behavioural capabilities. Considering scalability, protection, and the dynamic nature of IoT settings, the purpose is to expand a far better, more adaptable, and relatively secure way of authenticating IoT gadgets.
- The closing objective is to establish a validation gadget that boosts belief and dependability in IoT settings. Secure device onboarding, right of entry to control, and information transmission are the dreams of the recommended approach, which

mixes tool-precise features, historical information, and decentralized acceptance as true with mechanisms. Proving the effectiveness, scalability, and robustness of the counselled method in real IoT deployments, thorough simulation tests in several IoT contexts aid the intention.

The New Secure Validation Method for Evaluating the Legitimacy of IoT Devices is based on the literature review results in Sect. 2. Section 3 investigates the mathematical feasibility of the suggested approach; it is called Behavioural Decentralized Biometrics-Based Trust Score Authentication (BDB-TSA). The findings and analysis are detailed in Sect. 3, while Sect. 4 provides a brief overview and some recommendations.

2 Literature Survey

Strong security measures must be in place for the IoT because it is changing so quickly. Five separate contributions to IoT security are summarized in this review; they all deal with different problems and offer new ways to fix them.

The revolutionary anonymous authentication and key agreement system for heterogeneous IoT, introduced by Liu, J. et al., effortlessly transitions between certificateless cryptography (CLC) and public key infrastructure (CLC-PKI) [17]. It is lightweight and innovative. The technique guarantees secure communication, non-repudiation, user anonymity, and key agreement fairness. Proof of resilience against replay and DOS attacks has been found through security analysis. Based on the performance evaluation results, the scheme is a strong option for protecting sensitive data in various Internet of Things (IoT) applications due to its innovative features and lightweight design.

Using detection methods, deployment approaches, and validation procedures as criteria, the article by Khraisat A. et al. thoroughly evaluates modern IoT Intrusion Detection Systems (IDS) [18]. Gain valuable insights into the most popular approaches, datasets, and strategies for building intrusion detection systems. Along with offering a novel taxonomy to direct and unify efforts in improving IoT security, the paper describes detection capabilities, sorts IoT assaults, and lays out potential research obstacles.

Using cryptographic one-way hash functions, physically unclonable functions (PUF), and bitwise exclusive-OR (XOR) operations, Banerjee, S. et al. present a new protocol for anonymous user authentication in the Internet of Things (IoT) called novel lightweight anonymous user authentication protocol (NLAUAP) [19]. Formal verification using AVISPA and non-mathematical analysis is part of the thorough security analysis that the proposed scheme goes through. It shows resilience against different types of assaults. Comparative studies support its effectiveness in securing IoT environments, which display its superior overall performance in computing, communiqué fees, and security features. The recommended scheme's community overall performance characteristics are confirmed using a sensible demonstration of NS3 simulation.

A secure authentication approach (SAS) [20] with forward secrecy for Industrial Internet of Things (IIoT) structures is brought by Shuai M. et al. The recommended method fixes troubles with modern techniques, considerably with forward secrecy, using the Rabin cryptosystem and fending off password verification tables. The scheme is a well-rounded and powerful solution for sensible IIoT scenarios, as shown by using rigorous formal proofs and heuristic analyses, which grant the preferred purposeful and protection characteristics.

When it involves shielding IoT networks from replay, man-in-the-centre, and impersonation assaults, a brand new certificate-primarily based tool access management device (NC-DACS) [21] changed into delivered through Malani, S. et al. Strict safety features ensure the proposed scheme is proof against acknowledged assaults. These measures include formal verification using the AVISPA tool and real-or-random (ROR) version-based total analysis. By comparing it to relevant schemes, the counselled scheme offers a better compromise between protection, functionality, verbal exchange, and computing prices.

These creative contributions spotlight progress in Internet of Things protection, with every solution addressing unique issues. Compared to different techniques, the Revolutionary Anonymous Authentication and Key Agreement System (BDB-TSA) offers a thorough and efficient answer that provides incredible stability among efficiency, capability, and protection in the ever-changing Internet of Things (IoT) environment.

Proposed Method

A considerable obstacle inside the ever-converting IoT ecosystem is verifying the authenticity of linked gadgets. Because of the ever-converting nature of the IoT, limited resources, and comprehensive conversation protocols, traditional safety methods come up with problems. Behavioural Decentralized Biometrics-Based Trust Score Authentication (BDB-TSA) is a new technique supplied within the paper. A customizable and notably secure option for authenticating the authenticity of IoT devices is offered by using BDB-TSA, which leverages behavioural functions, blockchain, and least invasive cryptographic methods. It builds a stable basis for identifying devices and trustworthiness by combining tool-unique trends, historical facts, and decentralized trust mechanisms. Results from comprehensive simulation checks display that the proposed strategy is extra robust, scalable, and resilient to cyber threats in many IoT settings.

Blockchain technology makes the authentication procedure much more secure and reliable. It uses its inherent qualities of immutability, transparency, and scalability. A blockchain-based authentication system works on a distributed network of computers, each with access to the whole blockchain. By doing this, single points of failure are removed, and system resilience is increased. Blockchain prevents identity theft and fraud by authenticating persons, devices, or systems using digital signatures and cryptographic keys. A private key is used to sign each transaction as it is sent to the network. The network nodes use the associated public key to verify the transaction. Blockchain's transparency guards against manipulation and guarantees data integrity. A transaction is added to a block and the blockchain after it has been verified. The hashing process confirms the transaction's legitimacy by ensuring the data hasn't been altered.

Figure 1 indicates an intensive framework for adequately verifying the legitimacy of Internet of Things (IoT) gadgets, which solves the essential problem of ensuring all IoT devices are true and truthful. The "Device Legitimacy Assessment Module" is crucial to this architecture, a key component that determines whether Internet of Things gadgets are legitimate. Begin by going through the "Authentication Mechanism". This module contains several factors that affirm the authenticity of devices. The number one feature of authentication is to confirm the device's identity and rule out the opportunity of an attacker seeking to breach the network. The next layer specifies a "Secure Communication Channel" because these gadgets and the community as an entire can exchange information securely. Although the tool is authenticated, this keeps communication private and protected from feasible monitoring or manipulation. Incorporating sturdy security standards strengthens records transmission integrity, laying the groundwork for IoT surroundings; this is both secure and reliable.

The framework uses "Blockchain Technology" to bolster authentication and hold an everlasting report of device interactions—blockchain facts, authentication credentials and transactions in an immutable and decentralized ledger that cannot be altered.

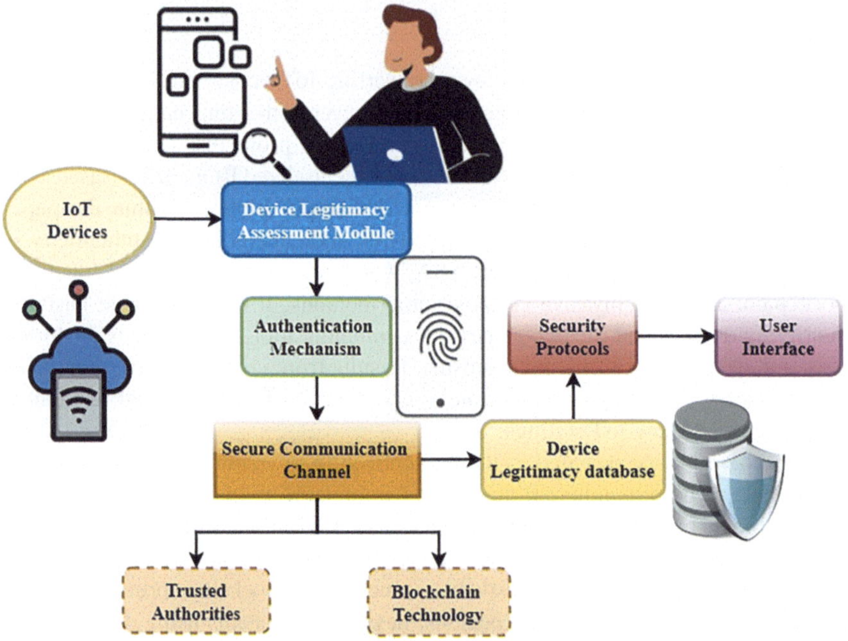

Fig. 1 Securely validating IoT device legitimacy

This makes the IoT network more secure, making it less difficult to listen to music and account for a person's moves. Trusted authorities are essential in this system because their operation ensures that gadgets are valid. These agencies may additionally maintain positions of belief in the IoT environment with manufacturers, regulatory corporations, or others. The system can confidently determine if gadgets are legitimate by drawing on that government's knowledge and validation skills.

The "Device Legitimacy Database" is a database that continues track of validated devices and the facts about them. The Device Legitimacy Evaluation Module uses this database as a fast reference factor to expedite validation and reduce reaction instances. Making it smoother to identify and put off malicious nodes strengthens the network as a whole. A "User Interface" is to be used as an extension to elevate a person's awareness and participation. To help clients sense greater price and confidence inside the IoT environment, this interface makes it smooth to peer whether or not their linked gadgets are legitimate. The provided validity popularity permits users to make educated selections, contributing to an extra safe and consumer-pleasant revel in the IoT.

Figure 1 keeps with an all-encompassing technique for adequately verifying the legitimacy of IoT gadgets. This framework lays a stable foundation for growing a reliable and secure Internet of Things (IoT) ecosystem by corporating authentication techniques, encrypted communique channels, blockchain technology, depended on authorities, and a significant data validity database. Improving the person reveal as a whole, the alternative consumer interface makes interactions with the Internet of Things (IoT) extra accessible and user-pleasant. Building a safe and resilient Internet of Things (IoT) infrastructure is the purpose of this all-encompassing technique, which addresses the complicated problems surrounding the legitimacy of IoT gadgets.

$$F = \frac{\beta . Q_{succ} . \left(\left(U^{\omega}_{beh} . U^{\sigma}_{phys} . U^{\varphi}_{crypt} \right) . U^{\alpha}_{env} \right)^{\delta}}{\gamma . \left(\left(\varepsilon . D^{\lambda}_{com} . D^{\mu}_{mem} . D^{L}_{net} . D^{\epsilon}_{energy} \right)^{Y} . S^{\omega}_{adapt} \right)^{\alpha} . S^{\eta}_{sec}} \quad (1)$$

Each variable is extraordinarily important within the efficiency (F) Eq. (1) for the BDB-TSA method, which stands for Behavioural Decentralized Biometrics-Based Trust Score Authentication. The chance of well-validating IoT gadgets such as layered accepted as true with components that include behavioural U^{ω}_{beh}, bodily U^{σ}_{phys}, environmental U^{φ}_{crypt}, and cryptographic U^{φ}_{crypt} agree with scores is denoted through Q_{succ}. Power adjustments ω, σ, α and φ are carried out to each accept as true with a score, incorporating complicated styles, physical developments, environmental situations, and diffused cryptographic dependability. The computational, reminiscence, community, and electricity sources wanted for validation are denoted as D^{λ}_{com}, D^{μ}_{mem}, D^{L}_{net}, and D^{ϵ}_{energy}, respectively. The intricate interdependencies between those sources are captured through a hyper energy adjustment Y. With energy change η that emphasizes vital security features and adaptive countermeasures, S^{ω}_{adapt} displays the system's adaptability to dynamic adjustments in IoT contexts and S^{η}_{sec} refers to the method's resistance against protection threats. These elements come collectively to offer an all-encompassing and context-aware version that can be used to assess how properly the BDB-TSA technique works within the complex world of Internet of Things device validation.

$$Acc(u) = \frac{\sum_{j=1}^{O} Q(D_j).\omega_{D_j}(u).J(D_j)}{\sum_{j=1}^{O} Q(D_j).\omega_{D_j}(u).J(D_j) + \sum_{k=1}^{N} Q(J_k).\omega_{J_k}(u).J(J_k)} \tag{2}$$

The depending time accuracy of the device for authentication in IoT environments is represented using Acc(u) in the accuracy Eq. (2). O and N are Included inside the variables, constituting the range of legitimate and invalid classifications. Dynamic weights $\omega_{D_j}(u)$ and $\omega_{J_k}(u)$ capture the changing relevance of these classifications through the years, whereas $Q(D_j)$ and $Q(J_k)$ imply the probability of legitimate and illegitimate classifications, respectively. When the corresponding categorization occurs, the indicator feature $J(D_j)$ and $J(J_k)$ assessment to 1, and otherwise, it evaluates to 0. An evolving probabilistic metric for accuracy that considers the sum of weighted chance for legitimate at the side of illegitimate classifications is provided through the overall equation. This metric adapts to changing IoT environments and accounts for the uncertainty related to behavioural traits and trust scores.

The system orchestrates seven essential components of the proposed Internet of Things data management system: sensors, hubs, controllers, clouds, databases, block chain, and users, as shown in Fig. 2. With these components working together, a solid foundation is created to help with the safe and effective management of data from the IoT. Essential to the system are users, who are authorized to access data from sensors, data analysis, and validation, among other IoT services. The IoT relies on hubs for networking, as its devices are known to have limited resources. While these devices can't run complex algorithms for cryptography, these devices are essential for building a secure connection to hubs and collecting data from the IoT. Intermediaries between the distributed network of IoT devices and the central system hubs collect data from several devices simultaneously. Hubs connect to the cloud and send encrypted messages using simple cryptographic methods like hash functions. Preserving the system's overall integrity, the assumption is made that IoT devices and hubs will have secure connections. Reliability and safety for IoT data is ensured by the controller, a single entity that stands at the core of the system and is responsible for numerous operations. Each connected hub has to have its precise timing coverage, which is the obligation of the Device Management Module. This new timing regulation makes recognizing Internet of Things devices easier and provides customized data transfer schedules to hubs. The scheduling process is susceptible to random changes, which are accommodated by the timing policy's margin of noise.

Once the distribution point has been located, the Database Maintenance Module can be activated. Before storing the data in the database, the controller computes its SHA256 hash. Returning the data index to the controller establishes a verifiable connection between the initial data and its exact placement in storage. To facilitate

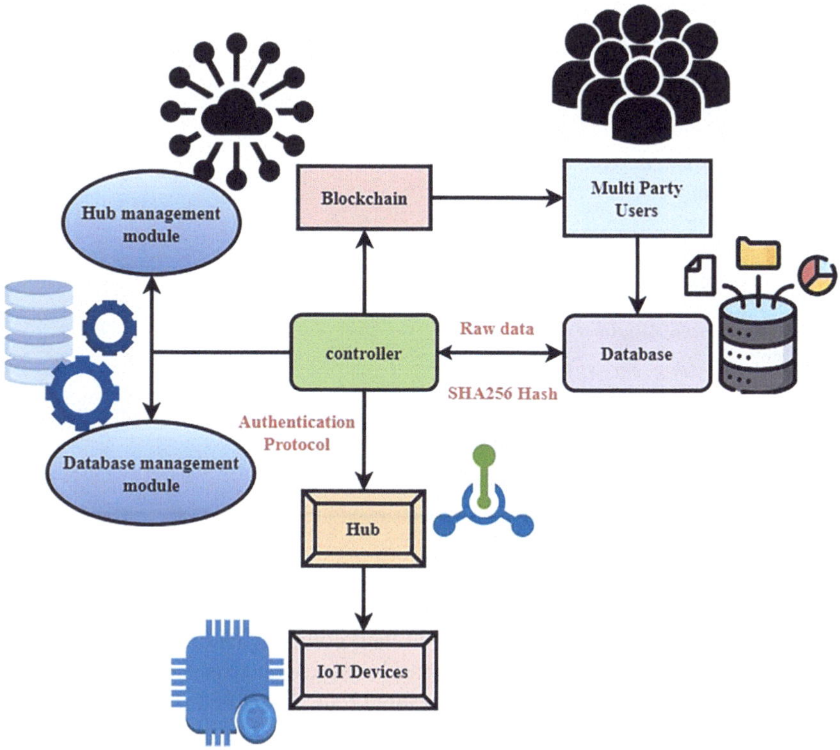

Fig. 2 Secure and effective IoT the management of data system

the generation of transactions, the Blockchain Administration Module presents the controller as a node on the blockchain. These transactions include the data's SHA256 hash, the controller's ID, and a database identification that includes the database's address and the data index. The blockchain keeps a permanent and verifiable record of all transactions to improve openness and security—Resource-Efficient Validation Scheme as shown in Algorithm 1.

Resource-Efficient Validation Scheme as shown in Algorithm 1

Input: m, B, t, f, q
Output: w_{tok}, F
Start:
1. Initial, m, B, t, f, q
2. For, j = 1; m
3. $m_{mem} \leftarrow msk$
4. $[e, f, o] \leftarrow g_2(msk, OQ$
5. $o_{core} = B_{jy}(H_j(s < U_j)$
6. $[w, jy] \leftarrow arg_{minimum}(dist)$
7. $[w_{tok}] \leftarrow g_2(msg, enc_j, o_j)$
8. $F_{uy} = g_3(dist, QM * cm)$
9. $F(o_{core}(j)] \leftarrow F(o_{core}(j) - F_{uy}$
10. *End*

End

The cloud stores authenticated data gathered from hubs in a database and hosts several controllers as a central repository. A public repository for sensor data, the database provides users with safe access to their company records. Users may trust the analytics and queries built on the database's indexed and hashed data. For reliable data audits, the blockchain is an indispensable tool. There is an immutable record of all system interactions, and users can check the validity of stored information by comparing blockchain transactions. Figure 2 shows a safe and all-encompassing framework for managing data from the Internet of Things. Utilizing cutting-edge timing policies, encryption algorithms, and blockchain technology guarantees that the data acquired is reliable, secure, and intact.

$$T = \frac{D.S\left(1 - f^{-\frac{U}{\sqrt{O}}}\right)}{\log(O+1)\frac{1}{Q}} \times \left(1 + \frac{G}{\prod_{j=1}^{N} H_j}\right) \times \left(1 + \frac{\sum_{k=1}^{L}\left(\frac{I_k}{\sqrt{M_k}}\right)}{\int_0^1 J(y)dy}\right) \quad (3)$$

Equation (3) for scalability uses the total hypercomplex Scalability Quotient, denoted as T, to quantify the scalability of the proposed method for validating IoT devices. Some of the variables are D, which captures the Comprehensive Computation Complexity, including biometrics, cryptography, and blockchain interpreting, and S, which represents the Robust Consistency Coefficient, representing the accuracy in trust score deduction. U stands for the Total Time Intricacy, which includes processing and communication times, and O is the Number of Internet of Things Devices in a network. The Q represents the Privacy Parameter, which emphasizes the features that provide privacy, while the $\log(O + 1)$ represents the logarithmic scaling impact caused by network resources. N represents the total amount of decentralized elements with H_j Serving as the Granularity Component for each, and G is the Flexibility Element that measures the system's adaptability. Where I_k is the

Heterogeneity Factor, and M_k is the Load Factor. The sum of all the dynamic impacts on scalability throughout a normalized range is contained in the Integration Functions $J(y)$. These factors combine to provide a complicated and nuanced evaluation of the suggested strategy for validating IoT devices in multifaceted and changing settings, taking scalability into account.

$$Q = \frac{\beta.\gamma}{\delta + \alpha.\epsilon - \frac{\vartheta}{\theta+\varphi} + \sqrt[3]{\Psi.\omega} + \int_0^1 YdY + \sum_{j=1}^{o} \eta_j \cdot \prod_{k=1}^{n} \mu_{jk}} \quad (4)$$

Equation (4) evaluates the BDB-TSA method's performance (Q) in validating the legitimacy of IoT devices. The β adaptable factor responds in real time to alterations in the IoT environment. The efficiency of combining behavioural traits, blockchain technology, and cryptographic processes is summed up by the synergy coefficient, denoted as γ. The δ represents the total computing burden, considering extracting features specific to the device, retrieving previous information, and implementing decentralized trust mechanisms. The α is the factor that optimizes resources, taking into consideration the fact that cryptographic operations are minimally intrusive. ϵ stands for the cyberattack security resilience parameter. The unpredictability in the behaviour of devices is captured by ϑ, which characterizes the system's entropy. θ measures how properly the authentication technique can manipulate a growing variety of gadgets, referred to as the scalability index. The reliability component is evaluated by φ, which determines how trustworthy the authentication system is. Ψ represents the innovation quotient, which is a degree of how extraordinary and adaptable the authentication method is. The self-gaining knowledge of the index, denoted by using ω, shows how properly the machine can modify its personnel. Continuous shifts and nice-tuning are denoted with the critical period Y. The η_j are factors that might change over time and affect overall performance, while μ_{jk} stands for nested product terms that seize complicated relationships. The quantity of nested interactions is denoted through n, while the quantity of dynamic additives is represented through o, both arbitrary constants. When put together, these components shape an all-encompassing assessment methodology for BDB-TSA in validating the legitimacy of IoT devices.

A new method of user identification called authentication makes use of trust scores, decentralized systems like blockchain, and the distinctive behavioural characteristics of people to improve reliability and confidentiality. This approach uses behavioral biometrics to continually authenticate users, such as mouse movements, typing patterns, or even use habits of the device. These distinctive and challenging-to-copy behavioural characteristics provide a strong means of verification. The use of blockchain technology accounts for the decentralized component. A blockchain-based solution reduces the number of single points of failure and increases system resilience by distributing the authentication process across many network nodes. The degree of trust in a user's identification may be measured using trust scores. Based on the user's behavioural biometrics' consistency over time, these scores are determined. When a user has a high trust score, it means that their current conduct is consistent with their past behaviour, which boosts confidence in their identification. Because

IoT devices often have limited resources, this strategy is especially appropriate for them. When behavioural biometrics are used, continuous, passive authentication is possible with minimal or no computing complexity. To further protect resources, the decentralized structure of blockchain eliminates the need for centralized servers.

IoT safety is of the utmost importance within the difficult IoT atmosphere. Data authenticity, secrecy, and integrity are becoming extra essential as the number of connected gadgets grows. Figure 3 reveals an all-encompassing, steady IoT atmosphere layout to bolster the basis of this rising technological frontier. The devices that make up the IoT are critical to this design. Unique styles of use, records transmission habits, and environmental interaction dynamics are encapsulated through those entities' particular behavioural tendencies. Each IoT entity is supported via a safe and diverse environment, further strengthened by tool-unique skills. A crucial part of the environment, the Trust Score Production tests the dependability of each Internet of Things tool. Behavioural traits and device-specific factors can be carefully tested in this observation. A device's reliability inside the ecosystem is quantified numerically by agreeing with the score. The decreased degree may result in more inspection or restricted entry, whereas a better score shows a more truthful device.

The structure uses Minimally Invasive Cryptographic Processes (MICPs) for enhanced safety. Cryptographic techniques that encode and decrypt statistics while preserving the tool's overall performance to a minimum are needed. The secure conversation is made possible with the assistance of lightweight cryptographic techniques that do not impact the working efficiency of the connected devices. Another vital aspect of the atmosphere is blockchain generation, which is famous for being decentralized and immune to tampering. A blockchain ledger facts all tool interactions and transactions, developing a transparent and unchangeable trail. By imposing

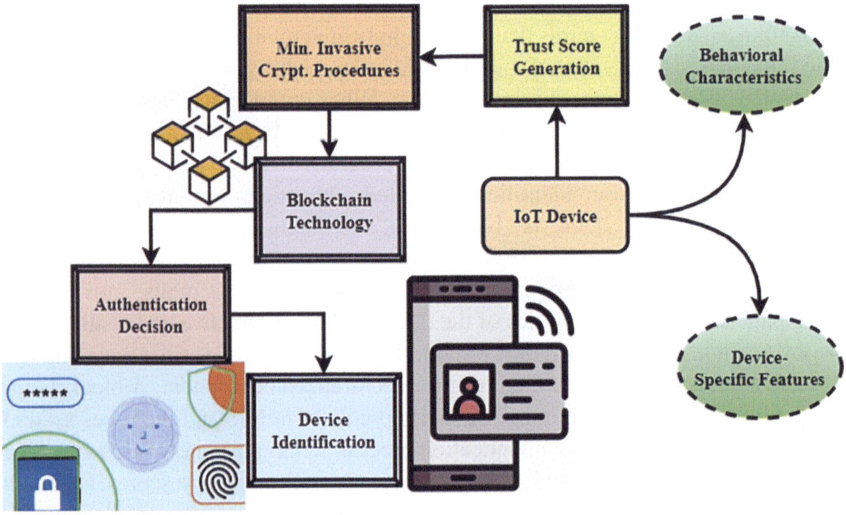

Fig. 3 IoT security architecture

measures including those, the IoT surroundings are more able to prevent unlawful rights of entry to and sell transparency and traceability. The Authentication Deciding module controls who may enter and how cash can be spent. Before authorizing or presenting an entry, it checks cryptographic signatures, blocks blockchain facts and accepts them as accurate with rankings. This cautious choice-making procedure limits participation inside the IoT environment to the handiest authenticated and sincere gadgets to reduce the possibility of unlawful access or dangerous interest.

Trustworthiness and Device Identification represent the last issue of the layout. Identification systems, such as unique identifiers and biometrics, allow for specific tool reputation. Devices contribute to safe and ever-changing surroundings through non-stop tracking and coffee evaluation of accept as true with ratings, which ensure that devices hold their integrity over the years. A comprehensive solution to the growing cyber protection threats in the IoT space is the covered IoT surroundings design proven in Fig. 3. Incorporating authentication techniques, cryptographic measurements, blockchain era, and behavioural analysis, this structure creates a robust framework that fosters acceptability and reliability among related gadgets. This layout offers a stable basis for the increasing IoT panorama, shielding the protection and soundness of the complete system. Secure and resilient digital destiny is confident by this sturdy IoT ecosystem design, which prioritizes acceptance as true with technology, cryptographic effectiveness, blockchain transparency, and strict authentication.

$$Trust\ score_{device} = \frac{\omega_1.F + \omega_2.D + \omega_3.C + \omega_4.E + \omega_5.\left(\sum_{j=1}^{o} I_j\right)^{\omega_7}}{\omega_6.Device_{capabilities}^{\omega_8} + \omega_9.\log_{10}(\omega_{10}.Additional\ security_{measures} + 1)} \quad (5)$$

The complete score that evaluates the overall protection and trustworthiness of an Internet of Things device is represented with the aid of Trust $\text{score}_{\text{device}}$ inside the dependability Eq. (5). The weights assigned to various factors are denoted by ω_1 thru ω_{10}, in which ω_1, ω_2, ω_3, ω_4, ω_5, ω_6, ω_7 to ω_8 imply the relative importance of various factors along with encryption energy, behavioural biometrics, the robustness of the block chain consensus mechanism, the dynamic protection aspect, the ancient security performance, the skills of the tool, the exponential period for historical overall performance, and the exponential period for device skills. The variable security factor, which captures the device's ability to adapt to new threats, is represented by E. When looking at the device's performance record throughout o, the historical security metrics for performance are represented by I_j. The IoT device's overall capabilities are represented as $Device_{capabilities}$. Additional security measures might have less of an effect beyond a certain point, as the logarithmic component introduces the concept of diminishing returns. The number or efficacy of the additional security measures that have been set in place $Additional\ security_{measures}$ affects the logarithmic term. Taking into account both the past and the decreasing returns of extra security measures, Eq. (5) provides a complex and detailed evaluation of the reliability of IoT devices.

$$BDS_{TSA} = \frac{\sum_{j=1}^{o}\left(\sqrt{\beta_j}.\log_{\gamma_j}(\delta_j)\right) + \alpha.f^{\epsilon/\vartheta} - \left(\frac{\lambda}{\theta} + \pi\right)}{\left|\cos\left(\frac{\sum_{j=1}^{O}\beta_j\cdot\gamma_j}{\delta_j+\alpha}\right) + \frac{\varphi}{\sqrt{\Psi}}\right|} \tag{6}$$

Equation (6) represents the Behavioural Decentralized Biometrics-Based Trust Score Authentication (BDB-TSA) for Internet of Things (IoT) devices, with each variable serving a distinct purpose. The square root and logarithmic operations highlight the complex interactions between δ_j, γ_j, and β_j, which together represent different behavioural features of IoT devices. The α, ϵ, and ϑ are related to the integration of blockchain technology, and their combined influence is underscored by an exponential operation. The parameters of device-specific features and previous information are represented by λ and θ. The value of O represents the total amount of behavioural traits that are taken into account. Adding mathematical constants, including f, π, and φ together with trigonometric and absolute value operations makes the suggested trust score authentication technique in the Internet of Things more difficult, which reflects its multifaceted character. The consider score technique of authentication becomes complicated by the Ψ.

The complex manner from tool acquisition to continual authentication is illustrated in Fig. 4, which exposes a dynamic interplay shape in the IoT surroundings. Dynamic digital surroundings necessitate a framework that can adapt to new threats and enhance the dependability of connected devices. When an Internet of Things (IoT) tool goes through the device boarding, the mixing process starts evolving. The machine uses its one-of-a-kind traits for behavioural authentication, which reveals its conduct of use, information transmission behaviour, and environmental interactions. This is the starting point for the Internet of Things surroundings. Each Internet of Things (IoT) tool is given a numerical estimate of its reliability through the Trust Score Calculation manner, which is important to the structure. This computation uses an aggregate of beyond statistics, tool-precise attributes and behavioural traits. A vital step towards ensuring the secure and decentralized preservation of critical information is the upward thrust of blockchain recording. The blockchain is used to document the produced belief score collectively with pertinent tool records. This improves the security of the Internet of Things (IoT) surroundings by stopping manipulation and illegal get right of entry to and by adding a further layer of transparency.

The following step is access management, which acts as a gatekeeper when agreeing with the rating. It takes the judge's position to determine whether gadgets are reliable enough to access community sources. Dynamic get admission to manage method is hooked up, responding to the extraordinary dependability tiers amongst devices. Higher trust ratings convert into greater right of entry to privileges. Priority and trustworthiness of transmitted facts are decided using acceptance as true with rating, which plays an essential component in the facts transmission procedure. The interconnected network benefits from efficient and secure data transmission because of this nuanced strategy, which prioritizes or implicitly trusts records from trusted

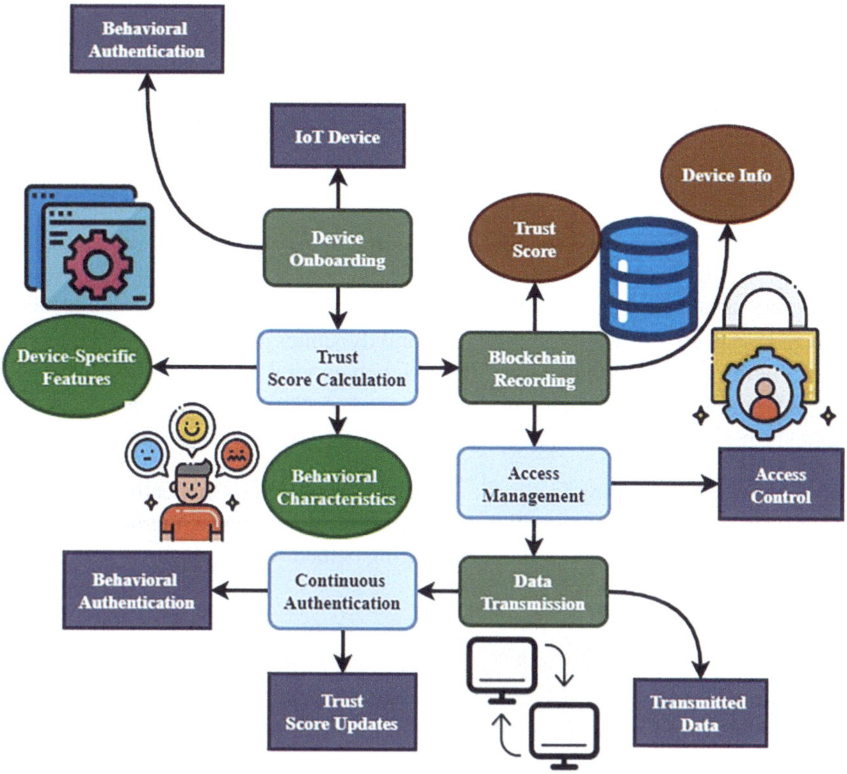

Fig. 4 Framework for dynamic interaction in IoT ecosystem

gadgets. The last step, Continuous Authentication, emphasizes the significance of non-stop monitoring. So that the device can regulate any modifications in a device's behaviour or attributes, behavioural authentication and trust rating updates appear periodically. Due to this iterative system, the Internet of Things (IoT) ecosystem can adapt to new protection threats effortlessly.

Figure 4 depicts an all-encompassing and bendy framework for guaranteeing dependability and protection within the good-sized Internet of Things (IoT) space. An ever-evolving framework for interplay is needed as the number of connected devices grows. This protocol lays a stable groundwork for the destiny of interconnected devices using coordinating responsibilities along with tool onboarding, agreeing with rating computation, blockchain taking pictures, getting admission to management, statistics shipping concerns, and ongoing authentication. A technique that promotes flexibility and safety while traversing the intricacies of the Internet of Things (IoT) surroundings. With the Internet of Things (IoT) landscape's rapid boom and diversity, Fig. 4's fluid communication structure exemplifies resilience and versatility. It ensures a trustworthy and dynamic digital future by addressing the current security and reliability issues and the future development of networked technology.

$$BDS_{TSA_{final}} = \frac{\sum_{j=1}^{o}\left(\frac{\partial g_j}{\partial u} + \int_0^U \frac{\partial^2 g_j}{\partial u^2} du\right).\omega_j.f^{\left(-\gamma.\sigma_j^2\right)}}{\sqrt{\sum_{k=1}^{n}\left(\int_0^U \frac{d_k}{du} du\right)^2}} \tag{7}$$

Behavioural Decentralized Biometrics-Based Trust Score Authentication $BDS_{TSA_{final}}$ is represented by Eq. (7). Equation (7) uses the n and o to represent the number of time dimensions and the duration of the behavioural data that is being analyzed, respectively. Capturing the temporal dynamics, the $\int_0^U \frac{d_k}{du} du$ is the square of integrating the first derivatives of the k th behavioural characteristic concerning time. It is possible to express the rate of change and acceleration of the j th behavioural feature as the product of its first and second derivatives, which are $\frac{\partial g_j}{\partial u} + \int_0^U \frac{\partial^2 g_j}{\partial u^2} du$. The γ controls the damping factor. $f^{\left(-\gamma.\sigma_j^2\right)}$, which is based on the feature's variance σ_j^2, and the weights ω_j measure the impact of each feature. This improved method of calculating the trust score considers changing behaviour patterns over time, using statistical traits and temporal derivatives.

$$Sim_{Level_{weighted}} = \frac{\sum_{l=1}^{u}\left(\frac{\int_0^U (\omega_l.UQ_l + (1-\omega_l).GO_l)du}{\int_0^U (\omega_l.UQ_l + (1-\omega_l).GO_l + GQ_l)du}\right)}{u} \tag{8}$$

In Eq. (8), the $Sim_{Level_{weighted}}$ is a modified simulation index that incorporates a time-weighted strategy to evaluate a method's efficacy in various dynamic Internet of Things (IoT) environments. Time is represented by the u, and different periods within the total period U are accounted for by the summation over l. Over the course of time, the weights ω_l modify the relative importance of true positives UQ_l, false negatives GO_l, and false positives GQ_l. False positives are cases mistakenly labelled as positive, while true negatives are correctly discovered situations. By including the time-dependent changes in success and error rates, the integrals give a holistic assessment of the method's efficacy throughout the period. Considering the dynamic nature of IoT environments, this improved simulation index affords an intensive review of the approach's efficacy.

To overcome the shortcomings of modern IoT protection frameworks, the encouraged method gives an intensive approach. Introducing a lightweight encrypting operation designed to permit decentralized ecosystems inside the Internet of Things (IoT) alleviates the problem of decentralization, one of the primary boundaries to be addressed. The counselled approach evaluates power factors simultaneously with the improvement of mystery keys, compared with conventional strategies that frequently run into issues because of complex encryption strategies. This novel technique ensures a more robust and long-term security mechanism while reducing resource-in-depth encryption's limitations. The advent of a targeted IoT structure methodically resolves the problem of unstructured communique. An extra organized and managed communique strategy is promoted by this structure, which organizes the association of IoT gadgets, including their interplay with a gateway node. The overall robustness and efficacy of the Internet of Things (IoT) system are stronger through this architectural development.

Implementing a safe modelling shape for nearby in addition to worldwide IoT, the cautioned methodology addresses the issue of biased identification regulations. It is set up that there is a distinction between the neighbourhood and the worldwide Internet of Things (IoT). The former refers to a network that supports software, while the one that follows is a centralized community that helps several diverse applications. The recommended method allows seamless interaction in any such heterogeneous setting using a domain-based total conversation device. In this setup, a gateway node facilitates communication between several local IoT devices, which forms a hierarchical and ordered network structure; each device is referred to as a domain. The first step in the modelling process is a basic description of the local IoT actors, such as servers, target Internet of Things (IoT) gadgets, and sensors.

Similarly, all data centres are portrayed as a component in the worldwide Internet of Things environment. A novel authentication method is introduced to ensure safe interaction between these parties. The technique guarantees a strong and dynamic authentication procedure by incorporating challenge and response-based approaches. Analytical methods are utilized in the study process for the proposed system, as shown in Fig. 5.

The next step in improving the authentication framework's security is to develop an algorithm that guarantees forward and reverse secrecy, according to the recommended

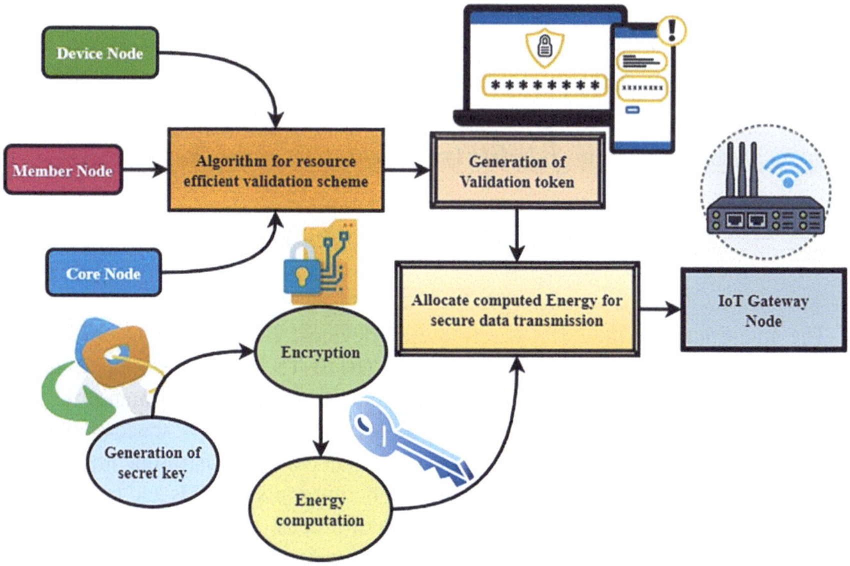

Fig. 5 Comprehensive methodology and algorithm design for IoT security

system. The technique uses cipher-text-based encryption with a novel energy computational approach to improve authentication while maintaining resource consumption. This model's unique selling point is its validation method, designed to provide increased security coverage by decentralizing the mitigation of dynamic risks in large-scale IoT systems. Figure 5 illustrates the suggested methodology and algorithmic design, which shows a new and systematic way to improve the system's resilience and efficiency by solving essential problems with Internet of Things security. Figure 5 details the proposed algorithm design and methodology, which incorporates new authentication and validation procedures for dynamic threat reduction in large-scale IoT environments and addresses decentralization, complex encryption limitations, and unstructured communication.

$$Sec\ level_{dynamic} = \log_2\left(\prod_{l=1}^{L}\left(\sum_{j=1}^{O}\left(\frac{\partial D_{jl}}{\partial u}.S_{jl}.Q_{jl}.\int_{0}^{U}\frac{\partial^2 D_{jl}}{\partial u^2}.\frac{\partial S_{jl}}{\partial u}.\frac{\partial Q_{jl}}{\partial u}.f^{-\alpha(\Delta u)^3}du\right)\right)\right) \quad (9)$$

The dynamic security level is represented by $Sec\ level_{dynamic}$ in Eq. (9). In terms of cryptographic complexity, resistance to assaults, and adaptability, the first-order time derivatives are represented by $\frac{\partial D_{jl}}{\partial u}$, $\frac{\partial S_{jl}}{\partial u}$, and $\frac{\partial Q_{jl}}{\partial u}$, respectively. The acceleration of cryptographic complexity is captured by the second-order time derivative, denoted as $\frac{\partial^2 D_{jl}}{\partial u^2}$. For each occurrence in a given time interval, S_{jl} and Q_{jl} stand for the resistance to attack and adaptation factors, respectively, for the j th instance. Over the period (U), where the rate of decay is controlled by α, the integral term computes the cumulative influence of second-order derivative and a cubic exponential decline $f^{-\alpha(\Delta u)^3}$. These dynamic elements accumulate across occurrences and the outer sum and product handle time intervals, and their accounting.

$$T_{BDB-TSA}(y,u) = \int N.G(E(y), J(y,u), U(y,u;\theta_{CD},\theta_D))dW_E + \int_{0}^{u} MT_{BDB-TSA}(y,t)dX_t \quad (10)$$

The mentioned method for securing IoT devices, Behavioural Decentralized Biometrics-Based Trust Score Authentication (BDB-TSA), relies on an advanced Eq. (10) with multiple variables. The manifold N represents an intangible area that includes many safety settings, and y is a factor within this space that specifies complex spatial interactions. The u represents time, illustrating how safety features evolve through the years. By adding inherent traits like IDs and behavioural peculiarities, tensor bundles like E(y) bring tool-precise information. The beyond importance of tool behaviour and protection occasions is captured by using the spatiotemporal vector package deal described as J(y, u). Parameterized by θ_{CD} and θ_D, the decentralized consider score is represented with the aid of the tensor bundles

$U(y, u; \theta_{CD}, \theta_D)$. A quantum field operator $T_{BDB-TSA}$ that evolves stochastically over the years through the Itô calculus and the non-commutative Dirac operators E(y) the non-commutative space reflects the advanced mathematical concepts used to evaluate and secure IoT gadgets, emphasizing the complex and ever-converting nature of the proposed BDB-TSA technique.

One progressive approach to verifying gadgets' legality in the vast Internet of Things (IoT) is the Behavioural Decentralized Biometrics-Based Trust Score Authentication (BDB-TSA) method. By combining behavioural tendencies with blockchain generation and less intrusive cryptographic approaches, BDB-TSA proves to be a stable, effective, and notably customizable answer that overcomes the shortcomings of previous strategies. Its tremendous overall performance, scalability, and robust resilience to cyber-attacks were shown with substantial simulation exams throughout diverse IoT environments. By setting up a new benchmark for reliability and belief inside the ever-changing global of related devices, BDB-TSA is well-positioned to convert real Internet of Things (IoT) deployments.

3 Results and Discussion

A progressive method for tackling essential troubles with IoT safety is the Novel Secure Validation Scheme for Assessing the Legitimacy of Internet of Things (IoT) Devices. The cautioned Behavioural Decentralized Biometrics-Based Trust Score Authentication (BDB-TSA) is compared to cutting-edge strategies in this thorough analysis that dives into essential performance indicators.

Impressive performance across important metrics is proven by evaluating the Novel Secure Validation Scheme for Assessing Device Legitimacy within the Internet of Things (IoT). Processing speed is one of the most critical factors to remember when designing IoT gadgets with limited assets. The recommended Behavioural Decentralized Biometrics-Based Trust Score Authentication (BDB-TSA) technique demonstrates high-quality performance in each region. Extensive simulation checks in numerous IoT situations show that the method can validate device legitimacy with minimum computing cost. Integrating blockchain generation into the validation procedure makes it more stable and transparent, leading to more reliable tool authentication. The scheme's flexibility is also tested by using the reality that it efficiently handles the ever-changing requirements of IoT environments. It supplies a reliable and effective solution to the troubles of tool onboarding, getting the right of entry to control, and encrypted information transfer. Efficiency testing confirms that the counselled validation approach outperforms rival strategies. It no longer has the most effective satisfied performance necessities, making it a robust contender for vast use in real Internet of Things (IoT) programs. Figure 6a indicates that the suggested gadget is more efficient than BDB-TSA, with an Efficiency Analysis showing a vast improvement of 99.4%. Figure 6b indicates that the cautioned gadget is more powerful than the LES alternative because the Efficiency Analysis shows that it is 85.6% more efficient.

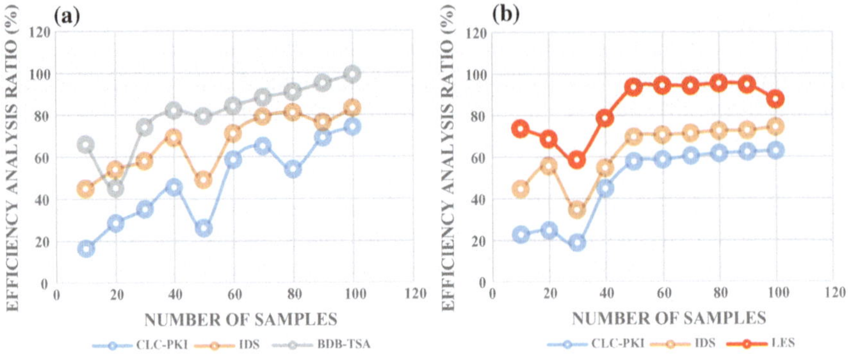

Fig. 6 **a** Efficiency analysis is compared with BDB-TSA. **b** Efficiency analysis is compared with LES

The dependability and accuracy of the Novel Secure Validation Scheme for Assessing Device Legitimacy within the Internet of Things (IoT) shows how well it verifies the legitimacy of linked gadgets. Using decentralized trust mechanisms and behavioural tendencies, the Behavioural Decentralized Biometrics-Based Trust Score Authentication (BDB-TSA) method achieves excessive accuracy in system identification. Improved authentication accuracy is a right-away result of using blockchain technology, which safeguards the validity method. By accomplishing thorough simulation reviews in various IoT environments, the scheme proves again to be more accurate than competing techniques. Offering a straightforward solution for tool legitimacy assessment, it accomplishes the tough project of handling the many distinct varieties of devices and the ever-changing nature of IoT environments. By analyzing the scheme's accuracy, one may see that it can distinguish among valid devices and that it's miles immune to protection threats. This method, known as the Novel Secure Validation Scheme, is ideal for making the Internet of Things extra straightforward and steady. The remarkable 96.8% overall performance compared to BDB-TSA within the Accuracy Analysis, as shown in Fig. 7a, demonstrates that the advised machine is more correct than BDB-TSA. In contrast, Fig. 7b suggests an assessment with LES, which yields an accuracy price of 58.8%. This highlights how much higher the recommended technique is than LES.

By studying its scalability, the Novel Secure Validation Scheme for Assessing Device Legitimacy in the Internet of Things (IoT) proves that it may keep up with the expanding surroundings of linked gadgets and remain successful even because it grows. As the variety and amount of Internet of Things (IoT) devices continue to develop, the Behavioural Decentralized Biometrics-Based Trust Score Authentication (BDB-TSA) technique shows remarkable scalability. Even in big Internet of Things (IoT) deployments, the approach might effectively affirm the legality of devices with little effect on computing resources, retaining overall performance. Integrating blockchain technology and decentralized consider mechanisms creates an allotted and stable framework for device authentication, which facilitates scalability.

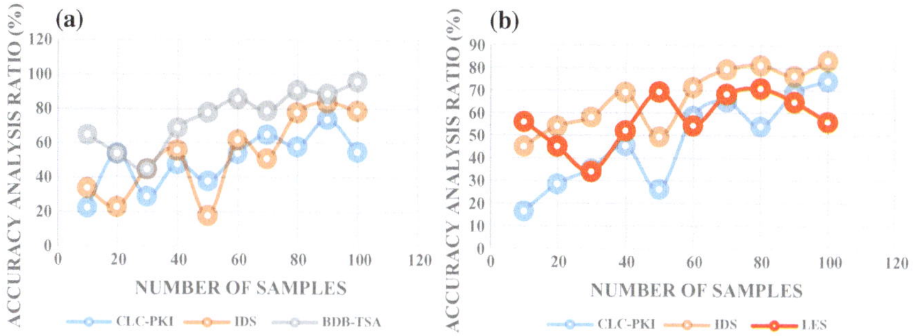

Fig. 7 **a** Accuracy analysis is compared with BDB-TSA. **b** Accuracy analysis is compared with LES

The scheme's scalability is confirmed via its capability to address the ever-converting and evolving nature of IoT environments. By conducting thorough simulation opinions in specific Internet of Things (IoT) environments, the research verifies that the Novel Secure Validation Scheme can scale to meet the developing needs of IoT tool validation whilst maintaining its performance and accuracy. Figure 8a suggests that the Scalability Analysis is extensively higher than BDB-TSA, with an efficiency rating of 94.6%. This highlights how the advised system outperforms BDB-TSA in phrases of scalability. Figure 8b compares with LES, which shows that the cautioned gadget is more scalable, with a score of 86.8%.

This novel stable validation scheme for comparing IoT device legitimacy has been well examined and confirmed to improve the safety of IoT environments. The BDB-TSA method (Behavioural Decentralized Biometrics-Based Trust Score Authentication) performs admirably in several distinct areas. With its rapid processing pace, the technique validates the legality of gadgets without setting a heavy computing burden on the device. Adding blockchain generation makes the validation technique more secure, improving the gadget's typical reliability. The scheme's robustness is further

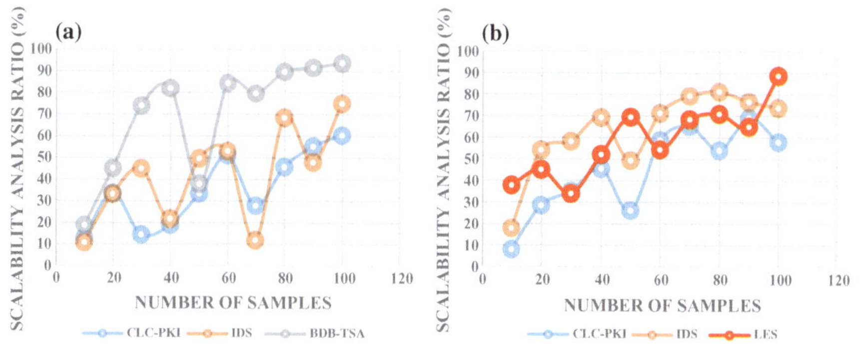

Fig. 8 **a** Scalability analysis is compared with BDB-TSA. **b** Scalability analysis is compared with LES

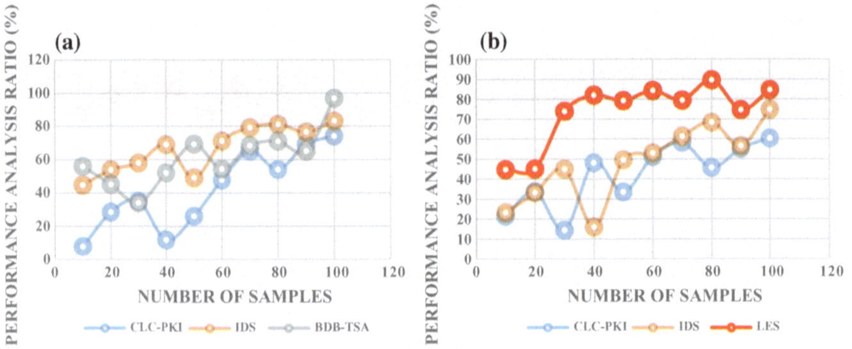

Fig. 9 **a** Performance analysis is compared with BDB-TSA. **b** Performance analysis is compared with LES

demonstrated by its flexibility in diverse IoT situations and its capacity to successfully control obstacles, along with limited resources and exceptional communication protocols. The studies confirm, through thorough simulation opinions, that compared to competing methodologies, the Novel Secure Validation Scheme plays higher in accuracy, scalability, and resilience to feasible security dangers. By supplying a strong foundation for tool authentication in practical IoT deployments, the scheme indicates promise to address the Internet of Things complex and ever-converting protection desires. The Performance Analysis, as shown in Fig. 9a, demonstrates an outstanding efficiency, surpassing BDB-TSA with an outstanding rating of 97.5%. This highlights the improved performance of the suggested system compared to BDB-TSA. In contrast, Fig. 9b

compares with LES, which yields a performance rating of 86.35%, demonstrating that the suggested system outperforms LES.

Examining the safety measures taken using the Novel Secure Validation Scheme to Determine the Legitimacy of IoT Devices demonstrates how properly it protects towards cybersecurity risks inside the IoT environment. Behavioural Decentralized Biometrics-Based Trust Score Authentication (BDB-TSA) is a technique that uses blockchain generation, decentralized belief mechanisms, and behavioural functions to reinforce device legitimacy evaluation. It is a multifaceted technique. Enhancing resistance towards tampering and unlawful access, the inclusion of blockchain guarantees the immutability of authentication data. To ensure the scheme's safety, sensitive information is protected at some point of validation by using the least invasive cryptographic strategies. There is much less danger of centralized vulnerabilities because decentralized trust mechanisms provide extra protection. After conducting thorough simulation exams in numerous Internet of Things (IoT) environments, the safety analysis confirmed that the cautioned scheme successfully protects against usual cybersecurity dangers, ensuring the privacy and authenticity of device authentication. Taken together, the security team looks at the proof that the Novel Secure Validation Scheme is a solid and straightforward answer, strengthening the safety of IoT settings and boosting belief in connected devices. Compared to BDB-TSA, the

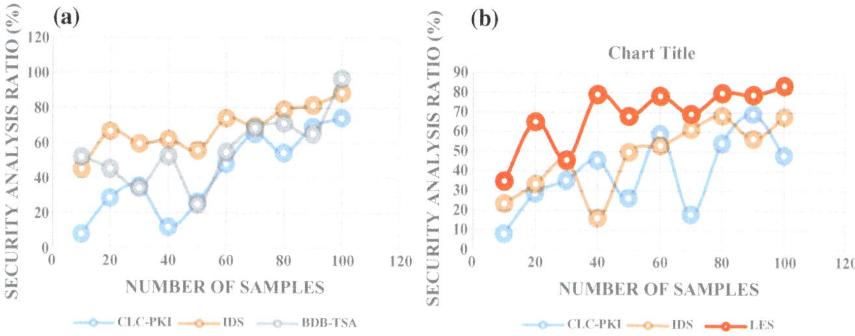

Fig. 10 **a** Security analysis is compared with BDB-TSA. **b** Security analysis is compared with LES

Security Analysis shows a strong overall performance in Fig. 10a, scoring a first-rate 96.8%. That the suggested technique is more steady than BDB-TSA is highlighted here. As shown in Fig. 10b, compared to LES, the recommended device has a good higher security rating of 85.6%, demonstrating its superiority.

The management and storage of this sensitive data could provide cause privacy problems since behavioral biometrics collects and analyzes user behavior patterns. IoT devices often have minimal processing and storage authority, which might not be enough for the complex computations required by blockchain technology.

With its attention on BDB-TSA, the Novel Secure Validation Scheme is an all-encompassing and pinnacle-notch solution that tackles the ever-changing Internet of Things (IoT) protection threats.

4 Conclusion

A new secure validation scheme for evaluating the legitimacy of IoT gadgets has been developed to address the severe problems resulting from the exponential boom of IoT devices. The improvement of the Internet of Things (IoT) has triggered an international shift in many industries; safeguarding those related devices from cybercriminals, maintaining personal data, and keeping confidence in IoT networks are essential. Because of their range and restrained assets, conventional safety strategies aren't constantly up to the task of protective IoT devices. It is acknowledged that a robust tool legitimacy evaluation scheme is crucial for preserving the dependability of IoT networks, and its importance is highlighted. To overcome the difficulties of authenticating the legitimacy of related gadgets, the recently proposed Behavioural Decentralized Biometrics-Based Trust Score Authentication (BDB-TSA) method suggests promise. A very safe, adaptive, and powerful approach to confirm the authenticity of IoT devices is offered by way of BDB-TSA. This technique uses accept as true with scores that might be derived from behavioural capabilities, blockchain generation and minimized cryptographic approaches. This technology establishes tool identity and dependable operation through device-specific traits, historical information, and decentralized agreement with mechanisms. After jogging the device through its paces in plenty of IoT environments, it became more effective, scalable, and resistant to cyber-attacks than competing techniques. In addition to proving the system works in concept, the simulation findings show that it can work in practice for Internet of Things deployments. With its foundation within the ground-breaking BDB-TSA approach, the Novel Secure Validation Scheme does more than most effectively solve the issues of figuring out the authenticity of IoT devices; it additionally sets a popular for future security paradigms within the ever-changing IoT environment. In an attempt to secure the ever-developing Internet of Things ecosystem, this study makes a widespread contribution to enhancing reliability, consideration, and safety in IoT contexts.

References

1. Omolara, A.E., Alabdulatif, A., Abiodun, O.I., Alawida, M., Alabdulatif, A., Arshad, H.: The Internet of things security: a survey encompassing unexplored areas and new insights. Comput. Secur. **112**, 102494 (2022)
2. Chen, S., Wen, H., Wu, J., Xu, A., Jiang, Y., Song, H., Chen, Y.: Radio frequency fingerprint-based intelligent mobile edge computing for Internet of things authentication. Sensors **19**(16), 3610 (2019)
3. Augusto-Gonzalez, J., Collen, A., Evangelatos, S., Anagnostopoulos, M., Spathoulas, G., Giannoutakis, K.M., Nijdam, N.A., et al.: From Internet of threats to Internet of things: a cyber security architecture for smart homes. In: 2019 IEEE 24th International Workshop on Computer Aided Modeling and Design of Communication Links and Networks (CAMAD), pp. 1–6. IEEE (2019)
4. Arun, M., Sivagami, S.M., Raja Vijay, T., Vignesh, G.: Experimental investigation on energy and exergy analysis of solar water heating system using zinc oxide-based nanofluid. Arab. J. Sci. Eng. **48**(3), 3977–3988 (2023)
5. Lu, Y., Wang, D., Obaidat, M.S., Vijayakumar, P.: Edge-assisted intelligent device authentication in cyber-physical systems. IEEE Internet Things J. **10**(4), 3057–3070 (2022)
6. El-Hajj, M., Fadlallah, A., Chamoun, M., Serhrouchni, A.: A survey of Internet of things (IoT) authentication schemes. Sensors **19**(5), 1141 (2019)
7. Mbarek, B., Ge, M., Pitner, T.: An efficient mutual authentication scheme for Internet of things. Int. Things **9**, 100160 (2020)
8. Nandy, T., Idris, M.Y.I.B., Noor, R.M., Kiah, L.M., Lun, L.S., Juma'at, N.B.A., Bhattacharyya, S., et al.: Review on security of Internet of things authentication mechanism. IEEE Access **7**, 151054–151089 (2019)
9. Lee, H., Kang, D., Ryu, J., Won, D., Kim, H., Lee, Y.: A three-factor anonymous user authentication scheme for Internet of things environments. J. Inf. Secur. Appl. **52**, 102494 (2020)
10. Arun, M., Barik, D., Sridhar, K.P., Dennison, M.S.: Thermal performance of a dimpled tube parabolic trough solar collector (PTSC) with SiO_2 nanofluid. Int. J. Photoenergy **28**, 2022 (2022)
11. Vinoth, R., Deborah, L.J., Vijayakumar, P., Kumar, N.: Secure multifactor authenticated key agreement scheme for industrial IoT. IEEE Internet Things J. **8**(5), 3801–3811 (2020)
12. Ali, I., Sabir, S., Ullah, Z.: Internet of Things Security, Device Authentication and Access Control: A Review. arXiv preprint arXiv:1901.07309 (2019)

13. Hussain, S., Ullah, S.S., Gumaei, A., Al-Rakhami, M., Ahmad, I., Arif, S.M.: A novel efficient certificateless signature scheme for the prevention of content poisoning attack in named data networking-based Internet of things. IEEE Access **9**, 40198–40215 (2021)
14. Kavianpour, S., Shanmugam, B., Azam, S., Zamani, M., Narayana Samy, G., De Boer, F.: A systematic literature review of authentication in Internet of things for heterogeneous devices. J. Comput. Netw. Commun. (2019)
15. Masud, M., Gaba, G.S., Choudhary, K., Hossain, M.S., Alhamid, M.F., Muhammad, G.: Lightweight and anonymity-preserving user authentication scheme for IoT-based healthcare. IEEE Internet Things J. **9**(4), 2649–2656 (2021)
16. Ankergård, S.F.J.J., Dushku, E., Dragoni, N.: State-of-the-art software-based remote attestation: opportunities and open issues for Internet of things. Sensors **21**(5), 1598 (2021)
17. Liu, J., Ren, A., Zhang, L., Sun, R., Du, X., Guizani, M.: A novel secure authentication scheme for heterogeneous Internet of things. In: ICC 2019–2019 IEEE International Conference on Communications (ICC), pp. 1–6. IEEE (2019)
18. Khraisat, A., Alazab, A.: A critical review of intrusion detection systems in the Internet of things: techniques, deployment strategy, validation strategy, attacks, public datasets and challenges. Cybersecurity **4**, 1–27 (2021)
19. Banerjee, S., Odelu, V., Das, A.K., Chattopadhyay, S., Rodrigues, J.J., Park, Y.: Physically secure lightweight anonymous user authentication protocol for Internet of things using physically unclonable functions. IEEE Access **7**, 85627–85644 (2019)
20. Shuai, M., Xiong, L., Wang, C., Yu, N.: A secure authentication scheme with forward secrecy for industrial Internet of things using Rabin cryptosystem. Comput. Commun. **160**, 215–227 (2020)
21. Malani, S., Srinivas, J., Das, A.K., Srinathan, K., Jo, M.: Certificate-based anonymous device access control scheme for IoT environment. IEEE Internet Things J. **6**(6), 9762–9773 (2019)

Ethics in AI: Balancing Innovation with Responsibility

Pooja Dehankar and Susanta Das

Abstract The swift progression of artificial intelligence (AI) technology offers unparalleled prospects as well as noteworthy ethical dilemmas. A strong ethical foundation is becoming more and more necessary as AI systems are incorporated into more areas of society, such as healthcare, banking, law enforcement, and everyday consumer products. It appears that current developments in AI come with significant challenges. This chapter emphasizes the significance of ethical frameworks in AI-enhanced contexts such as data-driven decision making, automation, adaptability and interactivity, intelligent assistance, and increased efficiency. This chapter explores the challenges associated with AI in detail. By striking a balance between scientific innovation and ethical ideals, we can harness the revolutionary potential of AI while minimizing its risks. AI has many potential applications in a wide range of industries, but as it has quickly gained traction, it has also brought up significant ethical issues that need to be carefully considered and addressed right away.

Keywords Artificial intelligence · Ethical concerns · Scientific innovation · Ethical frameworks · Accountability · Ethical dilemmas

1 Introduction

AI is a transformative technology that has the potential to drastically change a wide range of industries and have a significant impact on our daily lives. AI systems are becoming more and more common in many different industries, such as social

S. Das (✉)
School of Engineering, Ajeenkya DY Patil University, Charoli Bk. via Lohegaon, Pune, Maharashtra 412105, India
e-mail: susanta2024das@gmail.com; das29susanta@gmail.com

P. Dehankar
Ajeenkya DY Patil School of Engineering, Pune, Maharashtra 412105, India

S. Mahajan et al. (eds.), *Smart Systems: Engineering and Managing Information for Future Success*, Information Systems Engineering and Management 22,
https://doi.org/10.1007/978-3-031-76152-2_9

media platform algorithms, autonomous cars, healthcare diagnostics, and personalized recommendations. Given AI's continuous development and increasing societal integration, we need to address the ethical issues surrounding its study and application. AI systems are frequently able to learn and adapt more rapidly and accurately than individuals when presented with vast amounts of data. This can significantly increase productivity and innovation, even though it poses important ethical issues. Given their enormous power and possible repercussions, the principles, values, and policies that guide the creation and application of AI systems must be carefully examined. To align AI with human values and advance societal well-being, a careful balancing act between technological advancement and ethical responsibility is required. Transparency and explainability are important ethical factors. The intricacy and opacity of AI systems' decision-making and prediction processes can make them challenging to understand. Lack of transparency raises questions about responsibility, trust, and possibility of biased results [26].

In the area of technology, AI ethics is a rapidly developing field. Concerns concerning AI's effects on human existence, ethics, and morality are growing as its application across a range of industries grows. Artificial intelligence ethics can be viewed from two angles: human moral behavior is taken into consideration while developing, producing, utilizing, and caring for AI systems, and machine behavior is taken into consideration when discussing ethics [4]. The first viewpoint on artificial intelligence ethics deals with the moral dilemmas that come up during the development, usage, and handling of AI systems by humans (robot ethics). Concerns about data use in artificial intelligence systems and privacy, for instance, are being addressed. Furthermore, there exist concerns pertaining to the accountability of users of artificial intelligence systems for mistakes generated by the system, along with additional challenges associated with their utilization [9].

According to second viewpoint, moral concerns about machine behavior are addressed AI ethics. One concern, for instance, is making sure that racial or gender prejudice is not facilitated by artificial intelligence systems. Making sure AI systems don't have a detrimental effect on environment or society at large is another concern. Additionally, topic of potential singularities—i.e., instances in which artificial intelligence surpasses human intelligence and has an impact on human survival—is another ethical concern surrounding artificial intelligence [34].

In contemporary culture, ethics and law are closely intertwined, with several legal decisions originating from the interpretation of diverse ethical dilemmas. AI gives these problems a fresh angle. AI-powered systems are growing increasingly independent in terms of the kind of jobs they can do, the potential impact they may have on society, and the decreasing capacity of humans to comprehend, anticipate, and regulate their operations. The majority of people don't realize how autonomous these systems actually are. They are able to do actions that their creators did not intend for them to take and learn from their own experiences [25].

Since AI may mimic decision-making processes and behaviors, it is related to technologies that mimic human intellect. The majority of them are capable of handling challenging jobs on their own or with little assistance [28, 53, 54]. AI pertains to expert systems that continually learn through the acquisition of new data [5] through natural language processing [12], speech recognition [33], and/or machine vision [48].

Numerous industries are already reaping the rewards of artificial intelligence [16]. Numerous researchers have previously attested to the fact that AI systems are capable of automating repetitive jobs, including as data entry, online customer support, and invoice processing [45]. Expert systems possess rapid data analysis abilities, enabling them to optimize workflows in diverse contexts and make complex decisions more quickly and accurately than human beings. This has resulted in increased productivity and efficiencies across a range of industries [23, 35, 52].

The public sector, private businesses, and academic institutions have all released codes of ethics and standards for AI. Though it seems that everyone agrees that AI should be "ethical," there is disagreement on what "ethical AI" actually is and what ethical standards, best practices, and technical prerequisites are necessary to make it a reality. The findings show that there is a global convergence around five ethical principles (transparency, justice and fairness, non-maleficence, responsibility, and privacy), but there are also significant differences in the interpretation, significance, application, and issues, domains, and actors to which these principles apply [24].

It is imperative for companies to get ready for the impending wave of digital disruption that artificial intelligence is set to unleash. A few early adopting companies are already experiencing tangible benefits, thus it is more important than ever for the rest of the industry to move quickly with their digital transformations. AI has advantages, but also presents serious problems that affect businesses, developers, the government, and laborers. The workforce must be retrained to leverage AI rather than compete with it; nations and cities that are serious about becoming a global center for AI development must enter the global race to draw talent and capital in AI; advancements must be made in resolving the ethical, legal, and regulatory issues that could otherwise prevent AI from being used [10].

The Fairness, Accountability, and Transparency in ML forum created a Social Impact Statement for algorithms and Principles for Accountable Algorithms, while the World Economic Forum proposed nine ethical questions to ask AI systems on a global scale [49]. These organizations concentrate on creating trustworthy and responsible approaches to AI and digital innovation. Even though it's still unclear whether this goes beyond "ethics washing" in real life [21], there are now many guidelines and ethical AI principles in place [18], and idea of corporate digital responsibility (CDR [30]) is becoming increasingly popular in private sector.

This tendency is partly a reaction to well-known public disputes concerning data exploitation and the implementation of new legislation in the form of the EU General Data Protection Regulation (GDPR), which grants people more control over their personal data [42]. Following these advancements, concerns have surfaced regarding the necessity of ethical codes of practice and the degree to which regulatory compliance is adequate to guarantee ethical data practices [21]. Consequently, there is a growing understanding of the necessity of obtaining and preserving public support for the uses (and reuses) of individuals' data in order to create a "Social Licence" (SL) for practices that are carried out now and in future.

One of the main pillars of AI governance is accountability. This is caused by the assignment of duties (such forecasting or making decisions) to AI systems [37]. AI has two sides when it arises to solving business sustainability issues. The most challenging social and environmental issues that humans confront can be greatly improved by AI. However, the advances and efficiency brought forth by AI might also lead to new dangers, like automatic bias and ethical dilemmas. To handle the risks and sustainability issues posed by AI, businesses and governments should work together. A proactive regulatory framework bolstered by strict company rules and reports can facilitate the attainment of sustainable and accountable AI. To mitigate wide range of potential risks inherent in the deployment of AI, it is imperative to provide an acceptable level of technological neutrality and proportionality in the legislation. We can harness AI's potential for the good of society as a whole by closely evaluating these issues and working to strike a balance between innovation and responsibility. It will be feasible to speak of this technical advancement as a self-sufficient common good that is crucial to advancement of humanity [55].

Use of AI in decision-making raises a various ethical issues that need attention. Algorithm opacity prompts concerns about bias, accountability, and openness. AI-driven decision-making processes can be intricate and challenging to understand, making it difficult to comprehend how and why particular decisions are reached. This raises ethical issue about a lack of responsibility and transparency, particularly when these choices affect specific people or social groups. An important ethical dilemma is presented by bias in AI algorithms. Learning from past data, ML models have potential to reinforce preexisting biases within the AI system. To tackle this issue, careful analysis of the training set, algorithmic development, and continuous observation are needed to guarantee equity and reduce prejudice. Concerns concerning accountability and responsibility are also raised by the decision-making processes' growing reliance on AI. Determining the person who bears ultimate responsibility for actions made by AI systems presents a challenging ethical problem. To guarantee that people, organizations, and developers all bear some accountability for results of AI-driven decisions, a framework for accountability must be established. The impact of AI in decision-making on society as a whole is also subject to ethical concerns. Careful ethical analysis is needed to address issues including economic inequality, job displacement, and the possible concentration of power in the hands of a select few. Ensuring that AI helps society requires striking a balance between social responsibility and technological growth [40].

2 Digital Ethics

Some elements of digital ethics are nearly universal, such responsibility, transparency, beneficence, and justice; others, however, are more specific to particular kinds of organizations. A company looking to develop and implement new digital technologies, for instance, may find it especially crucial to balance the opportunity costs of passing up a potentially profitable new service, product, or innovation against the dangers of misuse and overuse, particularly in unregulated areas. According to Safavi and Kalis [47], a recent slowdown in the consumer uptake of digital health solutions due to mistrust in companies and a lack of understanding, highlights the significance of striking a balance between these principles in the regulatory adoption and consumer acceptance of modernizations. It also appears to be vital for new digital technology development.

Concerns about how AI may affect people and their social structures have given rise to the relatively new and developing topic of AI ethics. Conversely, the digital ethics, which covers related issues brought about by the creation and application of new digital technologies, includes AI ethics. Our fundamental societal and personal obligations, and our highest aspirations and values, are all under the purview of ethics. The attempt to rationally formulate and abide by principles or standards that reflect our fundamental societal and personal commitments, and our highest ideals and values is known as morality, and it is the foundation of ethics. Thus, digital ethics is the endeavor to regulate human behavior in creation and application of digital technology all together, and AI ethics is the endeavor to regulate human behavior in the creation and application of artificial automata, or machines, also known as computers, as a specific. This is achieved by logically establishing and adhering to principles or guidelines that represent our core values and commitments as individuals and as a society [20].

3 Trustworthy AI

Foundation of trustworthy AI is three fundamental pillars that support seven technological requirements that must be completed over the course of the system's whole life cycle: legality; ethics; and robustness, both in terms of technology and society. Achieving really trustworthy AI, however, requires a broader perspective that takes into account the reliability of all players and processes involved in the system's life cycle as well as an examination of earlier features from several angles. Four fundamental axes are considered in a more comprehensive view: the previously described pillars and requirements, a philosophical perspective on AI ethics, a risk-based approach to AI regulation, and global guidelines for the moral use and development of AI-based systems. The practical approach to implementing trustworthy AI systems analyzes requirements such as accountability, diversity, non-discrimination and many more. The idea that an accountable AI system is the outcome, and it is one of the most important concepts that may be achieved through auditing procedures, despite the difficulties presented by the application of regulatory sandboxes. A discussion of the opposing viewpoints recently published regarding the future of AI follows the multidisciplinary vision of reliable AI. A consensus between different points of view can only be reached through regulation, and reliable and accountable AI systems will be essential for the future and current state of our society [14].

4 AI Regulation

AI has been used with varying degrees of success in a variety of social contexts and areas, including as marketing, employment, finance, education, and law enforcement [17, 39]. In the comparatively early phases of AI development and use, the question of whether AI complies with certain ethical standards has emerged [3, 31]. Another important concern about how AI will be researched, used, and applied in the future is whether or not it can be governed by current laws [11, 29, 51]. The topic of governing AI was once limited to theoretical, technical, and academic discussions, but it has recently gained attention as governments and private businesses from significant geopolitical powers, such as the US, China, and the EU, have developed statements and policies on the subject of AI and ethics [41]. The latter also encompasses technology ethics, which is a subclass of AI ethics. In general, computer and engineering sciences—which are involved in the study and advancement of AI and ML—benefit from the self-reflection that AI ethics provides. Dynamics like particular technology development projects or the overall evolution of new technologies can be examined in this perspective. Likewise, a more static analysis can be used to look at the functionalities and causal mechanisms of specific technologies [44].

5 Ethical Concerns

Concerns regarding data security, privacy, autonomy, openness, and privacy are among the ethical issues raised by AI features. Transparency in ML is challenging because to its "black-box" nature, which limits human inspection and explanation of ML algorithms and makes it difficult for developers and customers to understand the inference process. Data security and privacy are crucial ethical considerations because the effectiveness of AI is heavily dependent on massive volumes of potentially sensitive data. These risks include malicious or abusive use, as well as information leakage and manipulation [22, 32]. Furthermore, AI advancements give systems a sense of purpose, accountability, and autonomy, this is particularly clear in healthcare robots. Morally challenging decisions must be made regarding AI systems' appropriate levels of autonomy, intentionality, and accountability.

In AI, determining accountability is about what happens when systems fail and produce undesirable outcomes. This issue, which is also referred to as "the problem of many hands," considers numerous factors, such as operational processes, programming codes, and input data. Everyone who develops, creates, implements, and uses AI systems in conjunction with humans is therefore liable. It is essential to develop ethical rules for AI in order to ensure that ethical behavior complies with ethical norms. This means that the ethical norms and notions that are currently in existence must be thoroughly understood by academics and practitioners. Additionally, individuals working on the design and implementation of AI systems need to be familiar with human rights regulations in order to prevent inadvertently breaching them [2, 27].

Issues regarding potential job losses are brought up by the increasing integration of robotics and automated systems in manufacturing, which raises issues related to automation and job displacement [13]. Moreover, human welfare is directly AI and other emerging technologies are becoming more widely available, and ethical considerations highlight the need to ensure that everyone in society has fair and equitable access to AI products and services [15].

6 Importance of Ethical Frameworks in AI

The framework's essential elements are security, privacy, accountability, and transparency. To build trust and allow for inspection, transparency means making sure AI algorithms and decision-making processes are understandable and explicable to stakeholders. Mechanisms for accountability are necessary to assign blame for AI-driven results and to make it easier for victims of injustice or harm to seek redress. Furthermore, the framework calls for the reduction of prejudice and discrimination against a range of demographic groups and highlights the significance of fairness in AI systems. Strong security procedures are necessary to prevent malicious exploitation and adversarial assaults, and privacy protection measures are considered critical to protecting people's sensitive data from unwanted access or usage. Conceptual framework seeks to enable developers, governments, and organizations to navigate difficult ethical terrain of AI development in IT systems by outlining best practices and ethical principles. In the end, implementing such a framework is essential to maximizing AI technology' revolutionary potential while maintaining core moral standards and social ideals [38].

Multiple fundamental theoretical frameworks that direct comprehension and action form foundation of ethical AI discourse. One important approach that emphasizes equality and fairness in algorithmic decision-making is Binns' application of Rawls' idea of justice to AI [6]. Particularly in light of Powers's examination of AI's moral and ethical decision-making, this strategy is consistent with Kant's deontological ethics. The understanding of the moral requirements and ethical issues inherent in AI development and application is made possible by these theoretical foundations [43].

The area of ethical AI has been greatly impacted by landmark research. Long-term existential concerns associated with AI, establishing a standard for thinking about the more general effects of sophisticated AI systems. Similarly, Russell's research emphasizes the connection of AI objectives with human ethical norms and presents practical concerns for ethical AI design. The way that scholars and practitioners tackle the ethical issues raised by AI has been influenced by these seminal publications [46].

These frameworks to tackle modern issues have evolved as a result of recent developments in AI ethics. A comprehensive method for comprehending the moral implications of IT, including AI, is offered by Floridi's framework of information ethics. Ethical questions raised by algorithms, particularly those pertaining to justice and bias, have also influenced the direction of recent debates and solutions in the arena of ethical AI [19].

Bias has far-reaching effects on equity and equality. It may lead to unfair employment procedures or loan approval procedures that penalize particular communities. Additionally, by maintaining preexisting prejudices found in training data, biased AI systems have the potential to promote discriminatory actions and stereotypes. This impedes the development of a more inclusive society in addition to undermining the ideas of justice and equality. Many methods are developed to minimize bias in AI systems in order to address these biases and guarantee equality and fairness. These include adjusting algorithms to lessen the impact of biased factors and employing a variety of training data that represents various opinions and demographics [7].

A methodical way to include transparency into AI development is through ethical AI frameworks. A number of groups and projects have established standards to handle moral issues like accountability, openness, and justice. The European Commission's Ethics Guidelines for Trustworthy AI and the IEEE Global Initiative on Ethics of Autonomous and Intelligent Systems are two notable examples [36]. These frameworks place a strong emphasis on values like accountability, justice, and transparency as the cornerstones of responsible AI development. For developers and organizations, ethical AI frameworks act as road maps, helping them to incorporate ethical issues into each phase of AI development. According to Vanclay et al. [50], this entails carrying out ethical impact assessments, making sure development teams are diverse, and interacting with stakeholders to comprehend society values and concerns.

The structures and rules in place now don't really address the issues of how to handle accountability issues. The precise application of newly proposed laws, like the AI Act, to corporate procedures and technological advancements is still unclear. A significant source of accountability dependencies is the growing incorporation of AI into products. The influence on individuals and society increases with system complexity. In addition, it is critical to provide a more transparent route for comprehending AI systems and the parties engaged in their development, application, and utilization when thinking about social responsibility and accountability. In case of accountability, much more work has to be done, both socially and technically [8].

7 Interdisciplinary Exploration of AI

Combining social science with AI has become a viable approach to solving difficult societal issues. This multidisciplinary strategy aims to maintain indigenous cultures in the area and advance sustainable development by utilizing the synergies between AI technologies and social science techniques. The critique of technological determinism, social constructivism, feminist viewpoints, and postcolonial theories are important themes that provide light on the power relations and moral dilemmas that arise while deploying AI. Through the integration of advanced AI techniques with conventional ecological knowledge, scientists may create customized approaches to tackle urgent issues like deforestation and land-use planning in native areas. It is imperative to exercise caution while navigating the ethical implications of AI deployment to uphold the rights and values of Indigenous populations. This means that to develop and apply AI solutions that are in line with community needs and goals, academics, policymakers, and local stakeholders must continue to communicate and work together. Furthermore, attaining significant and inclusive development outcomes depends on tackling structural problems including unequal access to technology and wider socioeconomic disparities. Thus, even if AI offers hope for good change, a comprehensive and inclusive strategy is necessary to realize its full potential for everyone's benefit [1].

8 Conclusion

AI has significantly advanced human civilization and altered conventional industrial techniques and ways of thinking. Although AI is a new technology, there is now no oversight mechanism and an imperfect policy structure. Risks associated with AI include environmental contamination, the expansion of the wealth gap, and privacy breaches. It also brings up moral questions like accountability, information, and human rights ethics. AI is evolving quickly to protect human rights and advance the positive social progress of society. In addition to promoting the development of AI ethics, we must fortify international cooperation and create sensible public policies.

One facet of the negative impact of AI technology is the ethical question surrounding it. Currently, many people believe that AI will lead to next productivity boom. The fundamental explanation for this is that AI has potential to significantly increase production, reduce need for human resources, enhance energy efficiency, decrease the cost of natural resources, advance social progress, and further human liberty. But everything has drawbacks and benefits, and science and technology are "double-edged swords" in general. Not only can artificial intelligence greatly facilitate human life, but it also poses a number of issues that need to be carefully considered to prevent issues and advance the advancement of human society.

In particular, issues like increased and uncontrollable military conflict, an abrupt and unavoidable rise in the unemployment rate, talent monopolies and overheating, the degradation of human spiritual and cultural life, a sharpening of the divide between the rich and the poor, social unrest, etc., could all be brought on by AI. Ethical issue raised by AI is important. It is a significant issue that human social growth cannot ignore and to which due consideration must be given in order to salvage the opportunity to rebuild the social ethical framework.

References

1. Affendi, R.: Interdisciplinary Exploration of AI and Social Science for Sustainable Development in Sabah, Malaysia. Malaysia (2024)
2. Aquino, Y.S.J.: Making decisions: bias in artificial intelligence and data-driven diagnostic tools. Aust. J. Gen. Pract. **52**(7), 439–442 (2023)
3. Arkin, R.C.: Ethical robots in warfare. IEEE Technol. Soc. Mag. **28**(1), 30–33 (2009)
4. Baker-Brunnbauer, J.: Management perspective of ethics in artificial intelligence. AI Ethics **1**(2), 173–181 (2021)
5. Berente, N., Gu, B., Recker, J., Santhanam, R.: Managing artificial intelligence. MIS Quart **45**(3) (2021)
6. Binns, R.: Fairness in machine learning: lessons from political philosophy. In: Conference on Fairness, Accountability and Transparency, pp. 149–159. PMLR (2018)
7. Bird, S., Kenthapadi, K., Kiciman, E., Mitchell, M.: Fairness-aware machine learning: practical challenges and lessons learned. In: Proceedings of the Twelfth ACM International Conference on Web Search and Data Mining, pp. 834–835 (2019)
8. Boch, A., Hohma, E., Trauth, R.: Towards an accountability framework for AI: ethical and legal considerations. Institute for Ethics in AI, Technical University of Munich, Munich, Germany (2022)
9. Boddington, P.: Towards a Code of Ethics for Artificial Intelligence, pp. 27–37. Springer, Cham (2017)
10. Bughin, J., Hazan, E., Sree Ramaswamy, P., DC, W., Chu, M.: Artificial intelligence the next digital frontier (2017)
11. Calo, R.: Robotics and the lessons of Cyberlaw. Calif. L. Rev. **103**, 513 (2015)
12. Carvalho, A., Levitt, A., Levitt, S., Khaddam, E., Benamati, J.: Off-the-shelf artificial intelligence technologies for sentiment and emotion analysis: a tutorial on using IBM natural language processing. Commun. Assoc. Inf. Syst. **44**(1), 43 (2019)
13. Dakshit, S., Dakshit, S., Khargonkar, N., Prabhakaran, B.: Bias analysis in healthcare time series (BAHT) decision support systems from meta data. J. Healthc. Inf. Res. **7**(2), 225–253 (2023)
14. Díaz-Rodríguez, N., Del Ser, J., Coeckelbergh, M., de Prado, M.L., Herrera-Viedma, E., Herrera, F.: Connecting the dots in trustworthy artificial intelligence: from AI principles, ethics, and key requirements to responsible AI systems and regulation. Inf. Fusion **99**, 101896 (2023)
15. Drukker, K., Chen, W., Gichoya, J., Gruszauskas, N., Kalpathy-Cramer, J., Koyejo, S., Myers, K., Sá, R.C., Sahiner, B., Whitney, H., Giger, M.: Toward fairness in artificial intelligence for medical image analysis: identification and mitigation of potential biases in the roadmap from data collection to model deployment. J. Med. Imag. **10**(6), 061104–061104 (2023)
16. Dwivedi, Y.K., Hughes, L., Ismagilova, E., Aarts, G., Coombs, C., Crick, T., Duan, Y., Dwivedi, R., Edwards, J., Eirug, A., Williams, M.D.: Artificial Intelligence (AI): multidisciplinary perspectives on emerging challenges, opportunities, and agenda for research, practice and policy. Int. J. Inf. Manage. **57**, 101994 (2021)
17. Ferguson, A.G.: The rise of big data policing: surveillance, race, and the future of law enforcement. In: The Rise of Big Data Policing. New York University Press (2017)
18. Fjeld, H., Achten, D., Feldman, K.: Principled Artificial Intelligence. A Map of Ethical and Rights-Based Approaches. Harvard University (2019)
19. Floridi, L., Taddeo, M.: What is data ethics? Philos. Trans. R. Soc. A Math. Phys. Eng. Sci. **374**(2083), 20160360 (2016)
20. Hanna, R., Kazim, E.: Philosophical foundations for digital ethics and AI Ethics: a dignitarian approach. AI Ethics **1**(4), 405–423 (2021)
21. Hasselbalch, G.: Making sense of data ethics. The powers behind the data ethics debate in European policymaking. Int. Policy Rev. **8**(2) (2019)
22. Hosseini, M., Resnik, D.B., Holmes, K.: The ethics of disclosing the use of artificial intelligence tools in writing scholarly manuscripts. Res. Ethics **19**(4), 449–465 (2023)

23. Javaid, M., Haleem, A., Singh, R.P., Suman, R.: Substantial capabilities of robotics in enhancing industry 4.0 implementation. Cogn. Robot. **1**, 58–75 (2021)
24. Jobin, A., Ienca, M., Vayena, E.: The global landscape of AI ethics guidelines. Nat. Mach. Intell. **1**(9), 389–399 (2019)
25. Karliuk, M.: Ethical and legal issues in artificial intelligence. In: International and Social Impacts of Artificial Intelligence Technologies, Working Paper, vol. 44 (2018)
26. Konidena, B.K., Malaiyappan, J.N.A., Tadimarri, A.: Ethical considerations in the development and deployment of AI systems. Eur. J. Technol. **8**(2), 41–53 (2024)
27. Kuhlberg, J.A., Headen, I., Ballard, E.A., Martin, D. Jr.: Advancing Community Engaged Approaches to Identifying Structural Drivers of Racial Bias in Health Diagnostic Algorithms. arXiv preprint arXiv:2305.13485 (2023)
28. LeCun, Y., Bengio, Y., Hinton, G.: Deep learning. Nature **521**(7553), 436–444 (2015)
29. Leenes, R., Lucivero, F.: Laws on robots, laws by robots, laws in robots: Regulating robot behaviour by design. Law Innov. Technol. **6**(2), 193–220 (2014)
30. Lobschat, L., Mueller, B., Eggers, F., Brandimarte, L., Diefenbach, S., Kroschke, M., Wirtz, J.: Corporate digital responsibility. J. Bus. Res. **122**, 875–888 (2021)
31. Mason, R.O.: Four ethical issues of the information age. In: Computer Ethics, pp. 41–48. Routledge (2017)
32. Merhi, M.I.: An evaluation of the critical success factors impacting artificial intelligence implementation. Int. J. Inf. Manage. **69**, 102545 (2023)
33. Narwani, K., Lin, H., Pirbhulal, S., Hassan, M.: Towards AI-enabled approach for Urdu text recognition: a legacy for Urdu image apprehension. IEEE Access (2022)
34. Nath, R., Sahu, V.: The problem of machine ethics in artificial intelligence. AI Soc. **35**, 103–111 (2020)
35. Ng, K.K., Chen, C.H., Lee, C.K., Jiao, J.R., Yang, Z.X.: A systematic literature review on intelligent automation: aligning concepts from theory, practice, and future perspectives. Adv. Eng. Inform. **47**, 101246 (2021)
36. Nikolinakos, N.T.: Ethical principles for trustworthy AI. In: EU policy and legal framework for artificial intelligence, robotics and related technologies—the AI Act, pp. 101–166. Springer International Publishing, Cham (2023)
37. Novelli, C., Taddeo, M., Floridi, L.: Accountability in artificial intelligence: what it is and how it works. AI Soc. 1–12 (2023)
38. Olorunfemi, O.L., Amoo, O.O., Atadoga, A., Fayayola, O.A., Abrahams, T.O., Shoetan, P.O.: Towards a conceptual framework for ethical AI development in IT systems. Comput. Sci. IT Res. J. **5**(3), 616–627 (2024)
39. O'Neil, C.: Weapons of math destruction: how big data increases inequality and threatens democracy. Crown (2017)
40. Osasona, F., Amoo, O.O., Atadoga, A., Abrahams, T.O., Farayola, O.A., Ayinla, B.S.: Reviewing the ethical implications of AI in decision making processes. Int. J. Manag. Entrepr. Res. **6**(2), 322–335 (2024)
41. Pichai, S.: AI at Google: our principles. Keyword **7**(2018), 1–3 (2018)
42. Politou, E., Alepis, E., Patsakis, C.: Forgetting personal data and revoking consent under the GDPR: challenges and proposed solutions. J. Cybersecur. **4**(1), tyy001 (2018)
43. Powers, T.M.: Prospects for a Kantian machine. IEEE Intell. Syst. **21**(4), 46–51 (2006)
44. Rahwan, I., Cebrian, M., Obradovich, N., Bongard, J., Bonnefon, J.F., Breazeal, C., Crandall, J.W., Christakis, N.A., Couzin, I.D., Jackson, M.O., Wellman, M.: Machine behaviour. Nature **568**(7753), 477–486 (2019)
45. Ribeiro, J., Lima, R., Eckhardt, T., Paiva, S.: Robotic process automation and artificial intelligence in industry 4.0—a literature review. Proc. Comput. Sci. **181**, 51–58 (2021)
46. Russell, S.: Artificial intelligence and the problem of control. In: Perspectives on Digital Humanism, p. 19 (2022)
47. Safavi, K., Kalis, B.: How can leaders make recent digital health gains last. Re-examining the accenture (2020)

48. Silva, R.L., Canciglieri Junior, O., Rudek, M.: A road map for planning-deploying machine vision artifacts in the context of industry 4.0. J. Ind. Prod. Eng. **39**(3), 167–180 (2022)
49. Stahl, B.C., Wright, D.: Ethics and privacy in AI and big data: implementing responsible research and innovation. IEEE Secur. Priv. **16**(3), 26–33 (2018)
50. Vanclay, F., Esteves, A.M., Aucamp, I., Franks, D.: Social impact assessment: guidance for assessing and managing the social impacts of projects (2015)
51. Wachter, S., Mittelstadt, B., Floridi, L.: Transparent, explainable, and accountable AI for robotics. Sci. Robot. **2**(6), eaan6080 (2017)
52. Wamba-Taguimdje, S.L., Wamba, S.F., Kamdjoug, J.R.K., Wanko, C.E.T.: Influence of artificial intelligence (AI) on firm performance: the business value of AI-based transformation projects. Bus. Process. Manag. J. **26**(7), 1893–1924 (2020)
53. Zhang, B., Zhu, J., Su, H.: Toward the third generation artificial intelligence. Sci. China Inf. Sci. **66**(2), 121101 (2023)
54. Zhang, C., Lu, Y.: Study on artificial intelligence: the state of the art and future prospects. J. Ind. Inf. Integr. **23**, 100224 (2021)
55. Zhao, J., Gómez Fariñas, B.: Artificial intelligence and sustainable decisions. Eur. Bus. Organ. Law Rev. **24**(1), 1–39 (2023)

Nanoenabled Materials Innovation and Optimization

Khan Dureshahwar, Kalam Sonali Sanjiv, and Buchake Dipak Gajanan

Abstract Nanotechnology has emerged as a transformative force in pharmaceutical research, offering innovative solutions to longstanding challenges in drug delivery, diagnostics, and tissue engineering. This chapter presents a comprehensive overview of recent advancements and optimization strategies in nano-enabled materials within the pharmaceutical field. Nano-enabled materials, including nanoparticles, liposomes, dendrimers, and nanofibers, exhibit unique physicochemical properties that can be tailored to enhance drug solubility, stability, and targeting specificity. Innovations in surface modification, functionalization, and ligand conjugation have enabled the optimization of nano-enabled materials for improved biocompatibility and therapeutic efficacy. Moreover, the integration of nanotechnology with other disciplines such as biotechnology and materials science has facilitated the development of multifunctional nanomaterials capable of synergistic drug delivery and imaging. However, challenges such as scalability, reproducibility, and regulatory considerations continue to impede the widespread clinical translation of nano-enabled materials. Addressing these challenges requires interdisciplinary collaboration and robust quality control measures to ensure the safety and efficacy of nano-enabled pharmaceutical products. By elucidating the latest trends and challenges in nano-enabled materials, this chapter aims to provide valuable insights for researchers, clinicians, and industry stakeholders striving to harness the full potential of nanotechnology in pharmaceutical applications. Nano-enabled materials in pharmaceuticals is expected to revolutionize drug delivery, diagnostics, and regenerative medicine, paving the way for more targeted, effective, and personalized healthcare interventions.

K. Dureshahwar (✉)
Maulana Azad Educational Trust's Y. B. Chavan College of Pharmacy, Aurangabad, Maharashtra, India
e-mail: dkmm31173@gmail.com

K. S. Sanjiv
Late Bhagirathi Yashwantrao Pathrikar College of Pharmacy, Aurangabad, Maharashtra, India

B. D. Gajanan
Encore Healthcare, Paithan, Aurangabad, Maharashtra, India

S. Mahajan et al. (eds.), *Smart Systems: Engineering and Managing Information for Future Success*, Information Systems Engineering and Management 22,
https://doi.org/10.1007/978-3-031-76152-2_10

Keywords Nanotechnology · Nano-enabled materials · Optimization strategies · Surface modification · Functionalization · Biocompatibility · Therapeutic efficacy · Interdisciplinary collaboration

1 Introduction

Nanotechnology is the manipulation of matter at the nanoscale, typically involving structures with dimensions between 1 and 100 nm. In pharmacy, nanotechnology plays a crucial role in designing, synthesizing, and utilizing materials and devices at the molecular or atomic level to address healthcare challenges. Its significance lies in its ability to revolutionize drug delivery, enhance therapeutic efficacy, improve patient compliance, and enable targeted treatments with reduced side effects. By harnessing the unique properties of nanomaterials, such as increased surface area, tunable reactivity, and improved biocompatibility, nanotechnology opens up new avenues for innovation in pharmaceutical research and development.

History

Nanotechnology's roots can be traced back to a visionary lecture by physicist Richard Feynman in 1959, where he discussed the possibility of manipulating individual atoms and molecules. However, the term "nanotechnology" was popularized by Eric Drexler in the 1980s through his book "Engines of Creation," which proposed the concept of molecular nanotechnology.

In 1981, the invention of the scanning tunneling microscope (STM) by Gerd Binnig and Heinrich Rohrer allowed scientists to visualize and manipulate individual atoms, marking a significant milestone in nanotechnology research. This breakthrough earned them the Nobel Prize in Physics in 1986.

Throughout the 1980s and 1990s, nanotechnology research expanded rapidly, with advancements in nanomaterial synthesis, characterization techniques, and theoretical understanding. In 1991, the discovery of carbon nanotubes by Sumio Iijima further fueled interest in the field.

The National Nanotechnology Initiative (NNI) was launched in the United States in 2000 to coordinate and fund nanotechnology research across various agencies. Since then, nanotechnology has grown into a multidisciplinary field with applications in electronics, materials science, medicine, and more.

Today, nanotechnology continues to evolve, driving innovation in diverse areas such as nanoelectronics, nanomedicine, nanomaterials, and Nano biotechnology. Its impact on science, technology, and society is profound, with potential implications for fields ranging from healthcare and energy to environmental sustainability and beyond [1].

Current Global Status of Nanotechnology in Pharmacy, Major Advancement and Trends

The current global status of nanotechnology in pharmacy reflects a dynamic and rapidly evolving landscape, marked by significant advancements and emerging trends. Here are some key highlights:

1. **Drug Delivery Systems**: Nanotechnology has revolutionized drug delivery by enabling precise control over drug release kinetics, targeting specific tissues or cells, and improving the solubility and stability of therapeutic agents. Nano formulations such as liposomes, polymeric nanoparticles, and dendrimers are being extensively explored for their potential to enhance the efficacy and safety of drugs.
2. **Personalized Medicine**: Nanotechnology is facilitating the development of personalized medicine approaches, where treatments are tailored to individual patients based on their genetic makeup, disease characteristics, and other factors. Nanoparticle-based diagnostics and therapeutics allow for precise disease detection, monitoring, and treatment optimization, leading to more effective and personalized healthcare interventions.
3. **Cancer Therapy**: Nanotechnology holds great promise for advancing cancer therapy by overcoming biological barriers, selectively targeting tumour cells, and minimizing off-target effects. Nanoparticle-based drug delivery systems, such as polymer-drug conjugates and lipid nanoparticles, are being explored for their ability to improve the efficacy of chemotherapy, reduce systemic toxicity, and overcome multidrug resistance in cancer cells [2].
4. **Bio imaging and Diagnostics**: Nanotechnology has revolutionized bio imaging techniques by enabling high-resolution imaging of biological structures at the nanoscale. Nanoparticle-based contrast agents, quantum dots, and other nanomaterials are being used for advanced imaging modalities such as magnetic resonance imaging (MRI), computed tomography (CT), and fluorescence microscopy. These advancements facilitate early disease detection, accurate diagnosis, and real-time monitoring of therapeutic responses.
5. **Regulatory Challenges and Safety Concerns**: As nanotechnology-based pharmaceutical products continue to emerge, regulatory agencies face challenges in ensuring their safety, efficacy, and quality. Issues such as nanoparticle toxicity, immunogenicity, and long-term biocompatibility require careful evaluation and regulatory oversight to mitigate potential risks and ensure patient safety.
6. **Global Collaboration and Investment**: Nanotechnology research in pharmacy is increasingly characterized by international collaboration and cross-disciplinary partnerships. Governments, academia, and industry stakeholders are investing heavily in nanotechnology research and infrastructure to accelerate innovation and translation of discoveries into clinical applications.

Overall, nanotechnology holds immense promise for transforming the field of pharmacy by enabling targeted and personalized therapeutic interventions, enhancing diagnostic capabilities, and addressing unmet medical needs. Continued research, collaboration, and regulatory diligence are essential to harnessing the full potential of nanotechnology for improving healthcare outcomes globally [3].

2 Nano Enabled Materials

Nanoenabled materials, also known as nanomaterials, are substances engineered at the nanoscale to exhibit unique physical, chemical, and biological properties that differ from those of their bulk counterparts. These materials have dimensions typically ranging from 1 to 100 nm and can be tailored with precision to achieve specific functionalities. Some common types of nanoenabled materials include (Fig. 1):

1. **Nanoparticles**: These are particles with at least one dimension in the nanoscale range. Nanoparticles can be composed of various materials such as metals (e.g., gold, silver), metal oxides (e.g., iron oxide, titanium dioxide), polymers, lipids, and carbon-based materials (e.g., carbon nanotubes, graphene). They are used in diverse applications including drug delivery, imaging, sensing, catalysis, and electronics.
2. **Nanostructured Surfaces:** Surfaces engineered at the nanoscale exhibit unique properties such as enhanced wettability, adhesion, and optical properties. Nanostructured surfaces find applications in self-cleaning coatings, anti-fogging materials, bioadhesives, and surface-enhanced spectroscopy.
3. **Nanocomposites**: These are materials composed of two or more components, where at least one component has nanoscale dimensions. Nanocomposites combine the properties of different materials to achieve synergistic effects, such as improved mechanical strength, thermal stability, or electrical conductivity. They are used in a wide range of applications including aerospace, automotive, construction, and electronics [4].

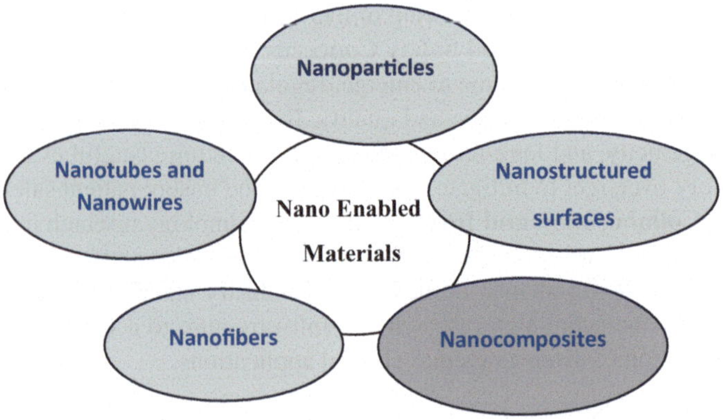

Fig. 1 Nano enabled materials

4. **Nanofibers**: These are fibers with diameters in the nanometer range. Nanofibers can be produced from various materials such as polymers, ceramics, and carbon-based materials. They possess high surface area-to-volume ratios and tunable porosity, making them suitable for applications such as filtration, tissue engineering, drug delivery, and sensors.
5. **Nanotubes and Nanowires**: These are cylindrical structures with nanoscale dimensions and high aspect ratios. Nanotubes can be composed of carbon (e.g., carbon nanotubes) or other materials (e.g., metal oxides, semiconductor materials), while nanowires are typically composed of semiconducting or metallic materials. They find applications in electronics, energy storage, photonics, and biomedical devices.

The unique properties of nanoenabled materials make them attractive for a wide range of applications across diverse industries, including healthcare, electronics, energy, environmental remediation, and consumer products. However, their synthesis, characterization, and safety assessment require careful attention to ensure their effective and responsible use in various applications [5].

What Nanoenabled Materials Are and How They Are Different from Traditional Materials

Nanoenabled materials, also known as nanomaterials, are substances engineered at the nanoscale to exhibit unique physical, chemical, and biological properties that differ from those of their bulk counterparts. Traditional materials, on the other hand, are typically composed of particles or structures with dimensions much larger than nanoscale [6].

Here are some key differences between nanoenabled materials and traditional materials:

1. **Size**: Nanoenabled materials have dimensions typically ranging from 1 to 100 nm, whereas traditional materials have larger dimensions at the macroscopic scale. This difference in size imparts unique properties to nanomaterials due to quantum effects and increased surface area-to-volume ratios.
2. **Properties**: Nanoenabled materials often exhibit properties that are distinct from those of bulk materials. These properties can include enhanced mechanical strength, optical transparency, electrical conductivity, catalytic activity, and magnetic behaviour. The unique properties of nanomaterials arise from quantum confinement effects, surface effects, and size-dependent phenomena [7].
3. **Surface Area**: Nanoenabled materials possess high surface area-to-volume ratios compared to traditional materials. This increased surface area allows for enhanced interactions with surrounding environments, such as improved adsorption, catalysis, and reactivity. It also enables nanomaterials to be functionalized with various molecules or ligands for specific applications.
4. **Behaviour**: At the nanoscale, materials may exhibit behaviour that differs from their bulk counterparts. For example, nanomaterials may display quantum confinement effects, where the electronic and optical properties are influenced by the confinement of charge carriers within nanoscale dimensions. Additionally,

nanomaterials may exhibit size-dependent melting points, mechanical properties, and magnetic behaviours [8].

5. **Applications**: Nanoenabled materials find applications across diverse fields, including electronics, healthcare, energy, environmental remediation, and consumer products. Their unique properties make them attractive for a wide range of applications, such as drug delivery, tissue engineering, sensors, catalysis, lightweight materials, and electronics.

Overall, nanoenabled materials represent a new class of materials with tailored properties and functionalities that hold immense promise for addressing technological challenges and advancing various industries. Their synthesis, characterization, and application require interdisciplinary approaches and careful consideration of their unique properties and potential implications [9].

Nanoenabled Materials Used in Pharmacy and Their Benefits

Certainly! Here are some examples of nanoenabled materials commonly used in pharmacy along with their benefits:

1. **Liposomes**: Liposomes are spherical vesicles composed of lipid bilayers that can encapsulate drugs within their aqueous core or lipid membrane. They are widely used in drug delivery due to their biocompatibility, ability to encapsulate hydrophobic and hydrophilic drugs, and potential for targeted delivery to specific tissues or cells. Liposomes protect drugs from degradation, enhance their solubility, and provide controlled release kinetics, thereby improving therapeutic efficacy and reducing side effects [10] (Fig. 2).

2. **Polymeric Nanoparticles**: Polymeric nanoparticles are colloidal particles composed of biodegradable polymers such as poly(lactic-co-glycolic acid) (PLGA), polyethylene glycol (PEG), or chitosan. They are used for controlled drug release, targeting specific tissues or cells, and improving the stability and bioavailability of drugs. Polymeric nanoparticles offer advantages such as tunable drug release kinetics, protection of drugs from enzymatic degradation, and reduced systemic toxicity.
3. **Dendrimers**: Dendrimers are highly branched, tree-like macromolecules with well-defined structures and precise control over size and surface functionality. They are used as drug carriers, imaging agents, and therapeutic agents in pharmacy due to their high drug-loading capacity, multiagency, and ability to penetrate biological barriers. Dendrimers enable targeted drug delivery, enhance drug solubility, and facilitate imaging and diagnosis of diseases [11].
4. **Carbon Nanotubes**: Carbon nanotubes (CNTs) are cylindrical structures composed of carbon atoms arranged in hexagonal lattice sheets. They possess unique mechanical, electrical, and thermal properties that make them attractive for drug delivery, bio sensing, and tissue engineering applications. CNTs can be functionalized with targeting ligands or therapeutic agents for targeted drug delivery to specific cells or tissues. They also serve as platforms for imaging, bio sensing, and controlled release of drugs [12].

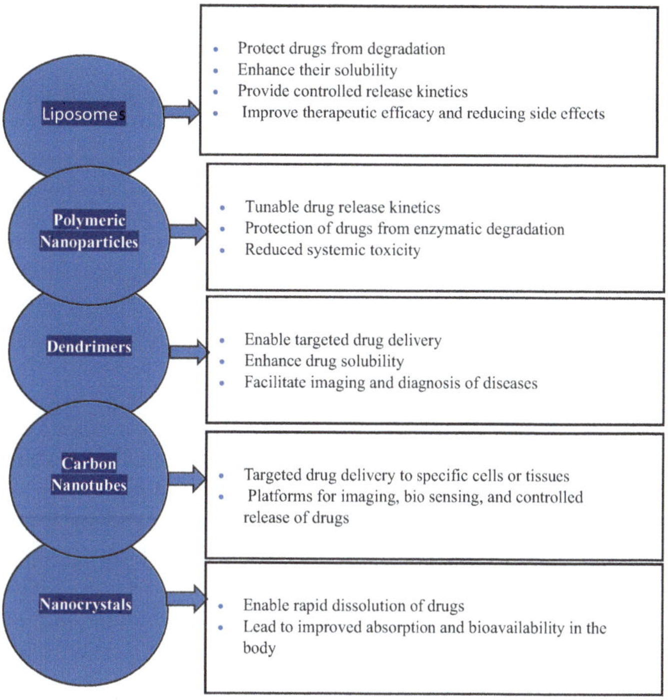

Fig. 2 Nanoenabled materials used in pharmacy

5. **Nanocrystals**: Nanocrystals are solid nanoparticles with crystalline structures and high surface area-to-volume ratios. They are used to enhance the solubility and bioavailability of poorly water-soluble drugs, thereby improving their therapeutic efficacy. Nanocrystal formulations enable rapid dissolution of drugs, leading to improved absorption and bioavailability in the body. They are particularly beneficial for drugs with low aqueous solubility, such as anticancer agents and antifungal drugs.

These nanoenabled materials offer significant advantages in pharmacy by addressing challenges such as poor drug solubility, low bioavailability, non-specific drug distribution, and systemic toxicity. Through precise control over drug release kinetics, targeting specificity, and biocompatibility, nanoenabled materials contribute to the development of safer, more effective, and patient-friendly pharmaceutical formulations [13, 14].

3 Integration of Nanotechnology with Other Discipline

The integration of nanotechnology with other disciplines is a multidisciplinary approach that involves collaboration between scientists, engineers, and researchers from various fields to address complex challenges and develop innovative solutions. Here are some examples of how nanotechnology is integrated with other disciplines:

1. **Medicine and Biology**: Nanotechnology intersects with medicine and biology to advance drug delivery, imaging, diagnostics, and regenerative medicine. Nanoparticle-based drug delivery systems enable targeted delivery of therapeutics to specific tissues or cells, improving efficacy and reducing side effects. Nanomaterials are also used for imaging and diagnosis of diseases, such as cancer, through techniques like nanoparticle-based contrast agents for MRI and fluorescence imaging. In regenerative medicine, nanotechnology is employed to design scaffolds and biomaterials for tissue engineering and organ transplantation [15].
2. **Chemistry**: Nanotechnology and chemistry intersect in the synthesis and functionalization of nanomaterials for various applications. Chemical methods are used to synthesize nanoparticles, Nano crystals, and Nano composites with controlled size, shape, and surface properties. Functionalization of nanomaterials with specific chemical groups allows for targeted drug delivery, bio sensing, and catalysis. Chemical principles are also applied to develop novel nanomaterial-based sensors and analytical techniques for environmental monitoring and medical diagnostics.
3. **Engineering**: Nanotechnology and engineering disciplines such as materials science, mechanical engineering, and electrical engineering converge to design and fabricate nanoscale devices, sensors, and systems. Engineering principles are applied to manipulate and characterize nanomaterials, as well as to develop nanofabrication techniques such as lithography, self-assembly, and template-assisted synthesis. Nanotechnology-enabled devices and sensors find applications in electronics, photonics, energy storage, and environmental sensing [16].
4. **Physics**: Nanotechnology draws on principles of physics to understand the behaviour of nanomaterials and exploit their unique properties. Quantum mechanics and solid-state physics provide insights into the electronic, optical, and magnetic properties of nanoscale materials. Physics-based techniques such as scanning probe microscopy, electron microscopy, and spectroscopy are used to characterize nanomaterials at the atomic and molecular level. Nanotechnology also intersects with fields like Nano photonics, plasmonics, and quantum computing to develop next-generation technologies with unprecedented capabilities [17].
5. **Environmental Science**: Nanotechnology is integrated with environmental science to develop sustainable solutions for pollution remediation, water purification, and energy conservation. Nanomaterials such as nanoparticles, nanofibers, and nanocomposites are used for wastewater treatment, air filtration, and soil remediation due to their high surface area, reactivity, and adsorption capacity.

Nanotechnology also enables the development of energy-efficient materials and devices for renewable energy generation and storage, such as nanomaterial-based photovoltaics, fuel cells, and batteries [18].

Overall, the integration of nanotechnology with other disciplines fosters cross-disciplinary collaboration and innovation, leading to transformative advancements with wide-ranging societal impacts. By leveraging the complementary expertise and perspectives of different disciplines, researchers can tackle complex challenges and harness the full potential of nanotechnology for addressing global needs [19].

Explore How Nanotechnology Intersects with Other Fields Such as Medicine, Biology, Chemistry, and Engineering

Nanotechnology intersects with multiple fields, including:

1. **Medicine**: In medicine, nanotechnology plays a pivotal role in drug delivery, imaging, diagnostics, and regenerative medicine. Nanoparticle-based drug delivery systems enable targeted delivery of therapeutics to specific cells or tissues, enhancing efficacy and reducing side effects. Nanomaterials are also used for imaging and diagnosis of diseases, such as cancer, through techniques like nanoparticle-based contrast agents for MRI and fluorescence imaging. In regenerative medicine, nanotechnology facilitates the design of scaffolds and biomaterials for tissue engineering and organ transplantation [20].
2. **Biology**: Nanotechnology provides powerful tools for studying biological systems at the nanoscale and manipulating biological molecules with precision. Nanoparticles and nanodevices are used as probes for imaging and monitoring biological processes in real time. Nanotechnology enables the development of biosensors and diagnostic devices for detecting biomarkers, pathogens, and toxins with high sensitivity and specificity. In addition, nanomaterials are employed for targeted delivery of nucleic acids, proteins, and other biomolecules for therapeutic applications [14].
3. **Chemistry**: Nanotechnology and chemistry intersect in the synthesis, functionalization, and characterization of nanomaterials with tailored properties and functionalities. Chemical methods are used to synthesize nanoparticles, nanocrystals, and nanocomposites with controlled size, shape, and surface properties. Functionalization of nanomaterials with specific chemical groups allows for targeted drug delivery, biosensing, and catalysis. Chemistry-based techniques are also applied to develop novel nanomaterial-based sensors and analytical methods for environmental monitoring and medical diagnostics [21].
4. **Engineering**: Nanotechnology and engineering disciplines such as materials science, mechanical engineering, and electrical engineering converge to design and fabricate nanoscale devices, sensors, and systems. Engineering principles are applied to manipulate and characterize nanomaterials, as well as to develop nanofabrication techniques such as lithography, self-assembly, and template-assisted synthesis. Nanotechnology-enabled devices and sensors find applications in electronics, photonics, energy storage, and environmental sensing [22].

The integration of nanotechnology with these fields fosters cross-disciplinary collaboration and innovation, leading to transformative advancements with wide-ranging societal impacts. By leveraging the complementary expertise and perspectives of different disciplines, researchers can tackle complex challenges and harness the full potential of nanotechnology for addressing global needs [23].

Collaborative Efforts and Interdisciplinary Research in Nanotechnology

Collaborative efforts and interdisciplinary research are essential components of nanotechnology, as they bring together experts from diverse fields to address complex challenges and drive innovation. Here's how collaborative and interdisciplinary approaches contribute to nanotechnology research:

1. **Integration of Expertise**: Nanotechnology encompasses a wide range of disciplines, including physics, chemistry, biology, materials science, engineering, and medicine. Collaborative efforts enable researchers with expertise in different areas to work together, leveraging their complementary knowledge and skills to tackle multifaceted problems. For example, physicists may contribute their understanding of quantum mechanics, chemists may provide expertise in synthesis and functionalization, and biologists may offer insights into biological interactions and mechanisms [24].
2. **Cross-Disciplinary Solutions**: Many nanotechnology challenges require cross-disciplinary solutions that draw on insights from multiple fields. For instance, the development of nanoscale drug delivery systems involves expertise in materials science, chemistry, pharmacology, and biomedical engineering. Collaborative research teams can design innovative solutions by integrating knowledge from diverse disciplines, leading to advancements in targeted drug delivery, imaging, diagnostics, and therapeutics [15].
3. **Innovation and Discovery**: Interdisciplinary research in nanotechnology often leads to unexpected discoveries and breakthroughs that drive innovation. By bringing together researchers with different perspectives and approaches, collaborative efforts foster creativity and exploration of new ideas. For example, interdisciplinary research teams may combine principles from biology and engineering to design biomimetic nanomaterials that mimic natural structures and functions, leading to novel applications in tissue engineering or drug delivery [25].
4. **Addressing Societal Challenges:** Nanotechnology has the potential to address pressing societal challenges such as healthcare, energy, environment, and sustainability. Collaborative research efforts focus on developing nanotechnology-based solutions to these challenges, such as improving access to clean water, enhancing renewable energy technologies, or advancing personalized medicine. Interdisciplinary approaches enable researchers to consider the broader societal implications of their work and develop ethically and socially responsible solutions [26].
5. **Education and Training:** Collaborative and interdisciplinary research in nanotechnology also extends to education and training. Interdisciplinary

programs and initiatives provide students and early-career researchers with opportunities to gain diverse skills and perspectives, preparing them for careers in academia, industry, and government. Collaborative research environments foster mentorship, collaboration, and knowledge exchange, nurturing the next generation of interdisciplinary scientists and engineers.

Overall, collaborative efforts and interdisciplinary research are crucial drivers of progress in nanotechnology, enabling scientists and engineers to address complex challenges, discover new phenomena, and develop innovative solutions with broad societal impact. By fostering collaboration across disciplines, researchers can unlock the full potential of nanotechnology to address global needs and improve quality of life [27].

4 Innovation in Nanotechnology

Innovations in nanotechnology have been transformative across various fields, driving advancements in materials science, electronics, medicine, energy, and environmental sustainability. Here are some notable innovations in nanotechnology:

1. **Drug Delivery Systems**: Nanotechnology has revolutionized drug delivery by enabling targeted and controlled release of therapeutics. Innovations such as nanoparticle-based drug carriers, liposomes, and polymeric nanoparticles allow for precise delivery of drugs to specific cells or tissues, improving therapeutic efficacy and reducing side effects [28].
2. **Nanostructured Materials**: Innovations in nanomaterial synthesis have led to the development of novel materials with tailored properties and functionalities. For example, carbon nanotubes, graphene, and quantum dots exhibit unique mechanical, electrical, and optical properties, enabling applications in electronics, sensors, and photonics.
3. **Nanomedicine**: Nanotechnology has transformed diagnostics, imaging, and therapeutics in medicine. Innovations such as nanoparticle-based contrast agents for medical imaging, targeted drug delivery systems, and nanoscale biosensors enable early disease detection, precise treatment, and personalized medicine approaches [29].
4. **Energy Storage and Conversion**: Nanotechnology has contributed to advancements in energy storage and conversion technologies. Innovations such as nanomaterial-based batteries, supercapacitors, and fuel cells offer higher energy densities, faster charging rates, and longer lifetimes, paving the way for renewable energy integration and electric vehicle adoption.
5. **Environmental Remediation**: Nanotechnology offers promising solutions for environmental remediation and pollution control. Innovations such as nanomaterial-based filters, catalysts, and sensors enable efficient removal of pollutants from air, water, and soil, addressing environmental challenges such as water contamination and air pollution [30].

6. **Nanoelectronics**: Innovations in nanoelectronics have enabled the development of smaller, faster, and more energy-efficient electronic devices. Nanoscale materials such as carbon nanotubes, graphene, and semiconductor nanowires enable advancements in transistors, memory devices, and sensors, driving progress in computing, communication, and electronics.
7. **Biomedical Devices and Implants**: Nanotechnology innovations have led to the development of biomedical devices and implants with enhanced biocompatibility and functionality. For example, nanocoatings and nanoscale surface modifications improve the performance and longevity of medical implants, while nanomaterial-based sensors enable real-time monitoring of physiological parameters.
8. **Targeted Drug Delivery Systems**: Recent innovations in nanotechnology have focused on developing targeted drug delivery systems that enable precise delivery of therapeutics to specific cells or tissues while minimizing systemic side effects. For example, researchers have developed nanoparticle-based drug carriers functionalized with targeting ligands that selectively bind to receptors overexpressed on cancer cells, allowing for site-specific drug delivery and enhanced therapeutic efficacy [31].
9. **Theranostic Nanoparticles**: Theranostic nanoparticles, which combine therapeutic and diagnostic capabilities, represent a significant innovation in nanomedicine. These nanoparticles enable simultaneous imaging of diseased tissues and delivery of therapeutic agents, facilitating early disease detection, monitoring of treatment response, and personalized medicine approaches. Theranostic nanoparticles can be tailored with imaging agents such as fluorescent dyes, magnetic nanoparticles, or quantum dots, along with therapeutic payloads such as drugs, nucleic acids, or proteins.
10. **Nanomedicines for Precision Medicine**: Advances in nanotechnology have enabled the development of nanomedicines tailored to individual patient characteristics, leading to the emergence of precision medicine approaches. By leveraging nanoscale drug delivery systems, researchers can customize treatment regimens based on a patient's genetic profile, disease subtype, or molecular signature, optimizing therapeutic outcomes and minimizing adverse effects. Precision nanomedicines hold promise for improving treatment outcomes in various diseases, including cancer, cardiovascular disorders, and neurodegenerative diseases.
11. **RNA-Based Nanotherapeutics:** Recent innovations in nanotechnology have expanded the repertoire of RNA-based nanotherapeutics for treating genetic diseases, infectious diseases, and cancer. Nanoparticle-based delivery systems enable efficient delivery of RNA molecules, such as small interfering RNA (siRNA), microRNA (miRNA), and messenger RNA (mRNA), to target cells, where they can modulate gene expression, regulate cellular processes, or induce therapeutic responses. RNA-based nanotherapeutics offer potential advantages in terms of specificity, potency, and versatility, opening up new avenues for disease intervention and treatment.

12. **Nanotechnology for Vaccine Delivery**: Nanotechnology has enabled the development of innovative vaccine delivery platforms that enhance vaccine efficacy, stability, and immunogenicity. Nanoparticle-based vaccine carriers, such as liposomes, polymeric nanoparticles, and virus-like particles, enable controlled release of antigens, adjuvants, and immunomodulatory agents, leading to enhanced immune responses and long-lasting protection against infectious diseases. Nanotechnology-based vaccine delivery systems also offer opportunities for developing next-generation vaccines targeting emerging pathogens and addressing vaccine delivery challenges, such as cold chain storage and distribution.

These innovations in nanotechnology demonstrate the transformative potential of nanoscale science and engineering in addressing societal challenges and driving technological progress. Continued research and development in nanotechnology promise further advancements with profound implications for diverse fields and industries.

Recent innovations and breakthroughs in nanotechnology have significantly impacted the field of pharmacy, offering novel solutions for drug delivery, diagnostics, imaging, and personalized medicine. Here are some notable advancements:

These recent innovations and breakthroughs in nanotechnology underscore its potential to revolutionize pharmacy by offering advanced drug delivery systems, personalized medicine approaches, and novel therapeutic modalities. Continued research and translation of nanotechnology-based solutions hold promise for addressing unmet medical needs, improving patient outcomes, and advancing the practice of pharmacy [32].

Emerging Technologies or Methodologies that Show Promise for Future Applications

Several emerging technologies and methodologies in nanotechnology hold promise for future applications across various fields, including medicine, electronics, energy, and environmental sustainability. Here are some notable examples:

1. **CRISPR-Cas9 Nanotechnology**: The combination of CRISPR-Cas9 gene editing technology with nanotechnology offers exciting possibilities for precise genome editing and targeted gene therapy. Nanoparticle-based delivery systems can efficiently deliver CRISPR components to target cells or tissues, enabling targeted genome modifications for treating genetic diseases, cancer, and infectious diseases. CRISPR-Cas9 nanotechnology holds promise for developing personalized medicine approaches and advancing gene therapy strategies.
2. **DNA Nanotechnology**: DNA nanotechnology involves the design and manipulation of DNA molecules to create programmable nanostructures and devices for various applications. DNA origami, in which DNA strands are folded into precise nanostructures, allows for the construction of nanoscale scaffolds, sensors, drug delivery vehicles, and molecular machines. DNA nanotechnology holds potential for advancing fields such as nanoelectronics, biosensing, drug delivery, and molecular computing [33].

3. **Nanopore Sequencing**: Nanopore sequencing is a next-generation sequencing technology that uses nanoscale pores to sequence DNA, RNA, or proteins in real time. Nanopore sequencers can directly analyze single molecules as they pass through a nanoscale pore, offering advantages such as rapid sequencing, long read lengths, and portability. Nanopore sequencing has applications in genomics, personalized medicine, infectious disease diagnosis, and environmental monitoring, with the potential to revolutionize the field of DNA sequencing.
4. **Metamaterials and Nanophotonics**: Metamaterials are engineered materials with unique electromagnetic properties not found in natural materials. By designing nanostructures with precise geometries and arrangements, researchers can manipulate light at the nanoscale, enabling applications such as cloaking devices, superlenses, and optical sensors. Nanophotonics, which combines nanotechnology with photonics, offers opportunities for developing ultra-compact photonic devices, high-speed communication systems, and efficient solar energy harvesting technologies [34].
5. **2D Materials and Beyond**: Two-dimensional (2D) materials such as graphene, transition metal dichalcogenides (TMDs), and black phosphorus exhibit extraordinary properties that make them promising candidates for future nanotechnology applications. These materials offer advantages such as high mechanical strength, exceptional electrical conductivity, and tunable optical properties. Researchers are exploring new methods for synthesizing and functionalizing 2D materials, as well as investigating their potential applications in electronics, photonics, sensors, and energy storage.
6. **Biohybrid Nanomaterials**: Biohybrid nanomaterials, which integrate biological components with synthetic nanomaterials, hold promise for developing functional materials with enhanced properties and functionalities. By harnessing the unique properties of biological molecules such as proteins, enzymes, and DNA, researchers can create hybrid nanomaterials with tailored structures and functions for applications in drug delivery, tissue engineering, biosensing, and environmental remediation [28].

These emerging technologies and methodologies demonstrate the potential of nanotechnology to drive innovation and address complex challenges in diverse fields. Continued research and development in these areas hold promise for advancing science, technology, and society in the coming years.

5 Pharmaceutical Applications of Nanotechnology

Pharmaceutical applications of nanotechnology encompass a wide range of innovative approaches that revolutionize drug delivery, diagnostics, imaging, and therapeutics. Here's a closer look at each application:

1. **Drug Delivery**: Nanotechnology enables precise control over drug delivery, improving efficacy and reducing side effects. Nanoparticle-based drug delivery systems, such as liposomes, polymeric nanoparticles, and dendrimers, enhance drug stability, solubility, and targeting. These nanocarriers can encapsulate drugs, peptides, or nucleic acids and deliver them to specific sites of action, such as tumors or inflammatory lesions, while minimizing systemic exposure.
2. **Diagnostics**: Nanotechnology enhances diagnostic capabilities by providing sensitive and specific detection methods for biomarkers, pathogens, and disease states. Nanoparticle-based biosensors, such as quantum dots, gold nanoparticles, and magnetic nanoparticles, enable rapid and multiplexed detection of analytes in biological samples. These biosensors offer advantages such as high sensitivity, low detection limits, and real-time monitoring, making them valuable tools for early disease detection and monitoring treatment response [35].
3. **Imaging**: Nanotechnology plays a crucial role in improving imaging modalities for disease diagnosis and treatment monitoring. Nanoparticle-based contrast agents, such as magnetic nanoparticles, gold nanorods, and carbon nanotubes, enhance the sensitivity and specificity of imaging techniques such as magnetic resonance imaging (MRI), computed tomography (CT), and optical imaging. These contrast agents provide better tissue delineation, improved resolution, and multiplexed imaging capabilities, facilitating early disease detection and precise localization of pathological lesions.
4. **Therapeutics**: Nanotechnology offers novel therapeutic approaches for treating various diseases, including cancer, infectious diseases, and neurological disorders. Nanoparticle-based drug formulations, such as liposomal doxorubicin, paclitaxel nanoparticles, and nanoparticle albumin-bound drugs, improve drug pharmacokinetics, biodistribution, and tumor targeting. Additionally, nanotechnology enables the development of targeted therapies, gene therapies, and immunotherapies that exploit nanoscale interactions to modulate cellular processes and enhance therapeutic outcomes.

These pharmaceutical applications of nanotechnology demonstrate its transformative potential to address unmet medical needs, improve patient outcomes, and advance the practice of medicine. As research and development in nanotechnology continue to evolve, we can expect further innovations and breakthroughs that will shape the future of healthcare [36].

Some Examples and Case Studies to Illustrate Each Application

Certainly! Here Are Specific Examples and Case Studies to Illustrate Each Application of Nanotechnology in Pharmacy

1. Drug Delivery

Example: Liposomal Doxorubicin (Doxil®)

Case Study: Doxil® is a liposomal formulation of doxorubicin used in the treatment of various cancers, including ovarian cancer, breast cancer, and multiple myeloma. Liposomes encapsulate doxorubicin, allowing for prolonged circulation in the bloodstream and selective accumulation in tumor tissues through the enhanced permeability and retention (EPR) effect. Clinical studies have demonstrated that Doxil® exhibits comparable or improved efficacy with reduced cardiotoxicity and myelosuppression compared to conventional doxorubicin formulations [37].

2. Diagnostics

Example: Quantum Dot-Based Biosensors

Case Study: Quantum dots (QDs) are semiconductor nanoparticles with unique optical properties that make them ideal for biosensing applications. Researchers have developed QD-based biosensors for detecting biomolecules such as proteins, nucleic acids, and viruses with high sensitivity and specificity. For example, a study published in ACS Nano demonstrated the use of QD-based biosensors for rapid and sensitive detection of influenza virus in clinical samples, offering potential applications in infectious disease diagnosis and surveillance [38].

3. Imaging

Example: Superparamagnetic Iron Oxide Nanoparticles (SPIONs) for MRI

Case Study: Ferumoxytol (Feraheme®) is an FDA-approved MRI contrast agent containing SPIONs coated with carboxymethyl dextran. Ferumoxytol is used for imaging blood vessels and detecting lesions in patients with chronic kidney disease. A study published in Radiology showed that Ferumoxytol-enhanced MRI provided superior vascular imaging compared to conventional gadolinium-based contrast agents, particularly in patients with impaired renal function. Ferumoxytol-enhanced MRI has applications in cardiovascular imaging, oncology, and inflammatory diseases [39].

4. Therapeutics

Example: Abraxane® (Nanoparticle Albumin-Bound Paclitaxel)

Case Study: Abraxane® is a nanoparticle albumin-bound (nab) paclitaxel formulation used in the treatment of breast cancer, non-small cell lung cancer, and pancreatic cancer. Abraxane® consists of paclitaxel nanoparticles bound to albumin, which enhances drug solubility, stability, and tumor targeting. Clinical studies have shown that Abraxane® exhibits improved response rates and survival outcomes compared to conventional paclitaxel formulations. In a phase III trial, Abraxane® demonstrated superior progression-free survival in patients with metastatic pancreatic cancer compared to gemcitabine, leading to FDA approval for this indication.

These examples and case studies highlight the diverse applications of nanotechnology in pharmacy, showcasing how nanotechnology enhances drug delivery, diagnostics, imaging, and therapeutics to improve patient outcomes and advance healthcare [40].

6 Conclusion

This chapter highlights the transformative potential of nanotechnology in material science. Through advancements in fabrication techniques and material design, nano-enabled materials offer unprecedented properties and functionalities. From enhancing mechanical strength to enabling targeted drug delivery, these materials are poised to revolutionize various industries. However, optimizing their synthesis and scaling up production remain crucial challenges. As research continues to push the boundaries of nanomaterials, interdisciplinary collaboration and rigorous experimentation will be essential for unlocking their full potential in addressing pressing global challenges and fostering sustainable development.

In conclusion, the chapter on Nano-Enabled Materials: Innovations and Optimization underscores the transformative potential of nanotechnology in various fields. It highlights the innovative approaches and methodologies employed in the design, synthesis, and optimization of nano-enabled materials. Through a comprehensive exploration of cutting-edge research and advancements, the chapter emphasizes the significance of nanomaterials in revolutionizing industries such as medicine, electronics, energy, and environmental remediation. Additionally, it underscores the importance of continuous innovation and optimization to unlock the full capabilities of nano-enabled materials, paving the way for enhanced performance, functionality, and sustainability. Overall, the chapter serves as a testament to the profound impact of nanotechnology in shaping the future of materials science and technology.

References

1. Nel, A., Xia, T., Mädler, L., Li, N.: Toxic potential of materials at the nanolevel. Science **311**(5761), 622–627 (2006). https://doi.org/10.1126/science.1114397
2. Zhang, L., Webster, T.J.: Nanotechnology and nanomaterials: promises for improved tissue regeneration. Nano Today **4**(1), 66–80 (2009). https://doi.org/10.1016/j.nantod.2008.10.014
3. Jain, K.K.: Nanomedicine: application of nanobiotechnology in medical practice. Med. Princ. Pract. **24**(1), 1–14 (2015). https://doi.org/10.1159/000366002
4. Kumar, C.S.S.R., Mohammad, F.: Magnetic nanomaterials for hyperthermia-based therapy and controlled drug delivery. Adv. Drug Deliv. Rev. **63**(9), 789–808 (2011). https://doi.org/10.1016/j.addr.2011.03.008
5. Zhang, X., Malhotra, S.V.: Synthesis, properties, and applications of carbon nanomaterials. In: Carbon Nanomaterials: Synthesis, Characterization, and Applications, pp. 1–32. CRC Press, Boca Raton (2015). https://doi.org/10.1201/b18789-2
6. Bhushan, B. (ed.): Handbook of Nanomaterials Properties. Springer, Berlin (2017). https://doi.org/10.1007/978-3-319-22389-1
7. Zhang, L., Webster, T.J.: Nanotechnology in Regenerative Medicine: Methods and Protocols. Springer, Berlin (2019). https://doi.org/10.1007/978-1-4939-9403-4
8. De, G., Kumar, A., Saxena, A.: Applications of Nanomaterials: Advances and Key Technologies. CRC Press, Boca Raton (2018). https://doi.org/10.1201/9781315151763
9. Nair, L.S., Laurencin, C.T. (eds.): Biomedical Applications of Nanotechnology. Wiley (2008). https://doi.org/10.1002/9780470379419
10. Torchilin, V.P. (ed.): Handbook of Nanobiomedical Research: Fundamentals, Applications and Recent Developments, vol. 4. World Scientific (2013). https://doi.org/10.1142/8582

11. Kumar, C.S.S.R.: Nanomaterials for Medical Diagnosis and Therapy. Wiley, London (2017). https://doi.org/10.1002/9781119361165
12. Lammers, T., Kiessling, F. (eds.): Nanomedicines: design, delivery and detection. R. Soc. Chem. (2019). https://doi.org/10.1039/9781788012336
13. Singh, R., Lillard, Jr J.W. (eds): Nanotechnology-Based Approaches for Targeting and Delivery of Drugs and Genes. Springer, Berlin (2019). https://doi.org/10.1007/978-3-030-04657-4
14. De Jong, W.H., Borm, P.J.: Drug delivery and nanoparticles: applications and hazards. Int. J. Nanomed. **3**(2), 133–149 (2008). https://doi.org/10.2147/IJN.S596
15. Ferrari, M. (ed.): Nanotechnology in Drug Delivery: Fundamentals, Design, and Applications. Springer (2005). https://doi.org/10.1007/0-387-26032-2
16. Cao, G. (ed.): Nanostructures for Drug Delivery. Elsevier, Amsterdam (2019). https://doi.org/10.1016/C2017-0-037667
17. Cao, G. (ed.): Nanostructures in Therapeutic Medicine. Elsevier, Amsterdam (2018). https://doi.org/10.1016/C2017-0-01011-9
18. Bhushan, B. (ed.): Springer Handbook of Nanotechnology. Springer, Berlin (2018). https://doi.org/10.1007/978-3-319-48933-9
19. Sargent, J.F. (ed.): Nanotechnology: business Applications and Commercialization. Springer, Berlin (2015). https://doi.org/10.1007/978-3-319-08437-6
20. Ferrari, M.: Nanotechnology in Drug Delivery: fundamentals, Design, and Applications. Springer, Berlin (2005). https://doi.org/10.1007/0-387-26032-2
21. Cao, G.: Nanostructures for Drug Delivery. Elsevier, Amsterdam (2019). https://doi.org/10.1016/C2017-0-03766-7
22. Bhushan, B.: Springer Handbook of Nanotechnology. Springer, Berlin (2018). https://doi.org/10.1007/978-3-319-48933-9
23. Zhang, L., Webster, T.J.: Nanotechnology in Regenerative Medicine: methods and Protocols. Springer, Berlin (2019). https://doi.org/10.1007/978-1-4939-94034
24. Kah, J.C., Wong, K.Y. (eds.): Nanotechnology for Nucleic Acid Delivery: Methods and Protocols. Springer, Berlin (2011). https://doi.org/10.1007/978-1-61779-029-8
25. Ruoff, R.S., Sander, L.M. (eds.): Nanotechnology research directions. In: IWGN Workshop Report, Vision for Nanotechnology R&D in the Next Decade. Springer, Berlin (2008). https://doi.org/10.1007/978-1-4020-8835-4
26. Roco, M.C., Bainbridge, W.S. (eds.): Converging Technologies for Improving Human Performance: Nanotechnology, Biotechnology, Information Technology and Cognitive Science. Springer, Berlin (2003). https://doi.org/10.1007/978-94-007-1044-8
27. Nel, A. E., Mädler, L., Velegol, D., Xia, T., Hoek, E.M., Somasundaran, P., et al.: Understanding biophysicochemical interactions at the Nano–bio interface. Nature Mater. **8**(7), 543–557 (2009). https://doi.org/10.1038/nmat2442
28. Blanco, E., Shen, H., Ferrari, M.: Principles of nanoparticle design for overcoming biological barriers to drug delivery. Nat. Biotechnol. **33**(9), 941–951 (2015). https://doi.org/10.1038/nbt.3330
29. Peer, D., Karp, J.M., Hong, S., Farokhzad, O.C., Margalit, R., Langer, R.: Nanocarriers as an emerging platform for cancer therapy. Nat. Nanotechnol. **2**(12), 751–760 (2007). https://doi.org/10.1038/nnano.2007.387
30. Anselmo, A.C., Mitragotri, S.: Nanoparticles in the clinic: an update. Bioeng. Transl. Med. **4**(3), e10143 (2019). https://doi.org/10.1002/btm2.10143
31. Bobo, D., Robinson, K.J., Islam, J., Thurecht, K.J., Corrie, S.R.: Nanoparticle-based medicines: a review of FDA-approved materials and clinical trials to date. Pharm. Res. **33**(10), 2373–2387 (2016). https://doi.org/10.1007/s11095-016-1958-5
32. Hua, S., de Matos, M.B.C., Metselaar, J.M., Storm, G.: Current trends and challenges in the clinical translation of nanoparticulate nanomedicines: pathways for translational development and commercialization. Front. Pharmacol. **9**, 790 (2018). https://doi.org/10.3389/fphar.2018.00790
33. Bae, Y.H., Park, K.: Targeted drug delivery to tumors: myths, reality and possibility. J. Control. Release **153**(3), 198–205 (2011). https://doi.org/10.1016/j.jconrel.2011.07.031

34. Jokerst, J.V., Gambhir, S.S., Dai, H.: Nanomaterials for in vivo imaging. Chem. Rev. **111**(3), 1776–1793 (2011). https://doi.org/10.1021/cr100702t
35. Jokerst, J.V., Lobovkina, T., Zare, R.N., Gambhir, S.S.: Nanoparticle PEGylation for imaging and therapy. Nanomedicine **6**(4), 715–728 (2011). https://doi.org/10.2217/nnm.11.19
36. Davis, M.E., Chen, Z., Shin, D.M.: Nanoparticle therapeutics: an emerging treatment modality for cancer. Nat. Rev. Drug Discovery **7**(9), 771–782 (2008). https://doi.org/10.1038/nrd2614
37. Doxil®: Barenholz, Y.: Doxil®—The first FDA-approved nano-drug: lessons learned. J. Control Release **160**(2), 117–134 (2012). https://doi.org/10.1016/j.jconrel.2012.03.020
38. Quantum Dot-Based Biosensors: Qian, J., Jiang, L., Cai, F., Wang, D., He, S., Hu, Q. et al: Quantum dot-based immunochromatography test strip for rapid, quantitative and sensitive detection of prostate-specific antigen. Biosens. Bioelectron. **24**(11), 3088–3094 (2008). https://doi.org/10.1016/j.bios.2009.03.025
39. Ferumoxytol (Feraheme®): Bashir, M.R., Bhatti, L., Marin, D., Nelson, R.C.: Emerging imaging technologies: ferumoxytol MR angiography. RadioGraphics **35**(2), 482–498 (2015). https://doi.org/10.1148/rg.352130136
40. Abraxane® (Nanoparticle Albumin-Bound Paclitaxel): Gradishar, W.J., Tjulandin, S., Davidson, N., Shaw, H., Desai, N., Bhar, P. et al: Phase III trial of nanoparticle albumin-bound paclitaxel compared with polyethylated castor oil–based paclitaxel in women with breast cancer. J. Clin. Oncol. **23**(31), 7794–7803 (2005). https://doi.org/10.1200/jco.2005.04.937

Governance Frameworks for Smart Systems

Harpreet Kaur Channi, Ramandeep Sandhu, Nimisha Singh, Deepika Ghai, Pradeep Kumar, and Gagandeep Singh Cheema

Abstract Effective governance frameworks are essential in the fast-growing world of smart systems, which include networked technology, Artificial Intelligence (AI), and the Internet of Things (IoT). Smart systems have unique governance difficulties, which this abstract explores in detail. For future success, engineering concepts and modern information management tactics are integrated. Governance is crucial to ethical, safe, and efficient smart systems, as the research explores their many facets. The study presents frameworks that smoothly integrate technology innovation with strong governance structures using engineering insights. These frameworks emphasize iteration to match smart systems' dynamic nature. The suggested information management frameworks include data integrity, privacy, and accessibility. The study examines innovative methods for managing smart system data, emphasizing information stewardship. The research also examines ways to make data use transparent and accountable to improve society. The focus is on how governance frameworks may help engineers, legislators, industry leaders, and the public collaborate. A holistic strategy that includes multiple stakeholders' interests and concerns promotes collaborative responsibility for smart system development and implementation, according to the research. As smart systems become essential to healthcare, transportation, and urban planning, the study proposes sector-specific governance frameworks. These

H. K. Channi
Department of Electrical Engineering, Chandigarh University, Gharuan, Mohali 140413, India

R. Sandhu · P. Kumar
School of Computer Science and Engineering, Lovely Professional University, Jalandhar, Punjab 144411, India

N. Singh
Department of Skill Development, Baba Farid College of Engineering and Technology, Bathinda, Punjab 151005, India

D. Ghai (✉)
School of Electronics and Electrical Engineering, Lovely Professional University, Phagwara, Punjab 144411, India
e-mail: money.ghai25@gmail.com

G. S. Cheema
Mittal School of Business, Lovely Professional University, Phagwara, Punjab 144411, India

S. Mahajan et al. (eds.), *Smart Systems: Engineering and Managing Information for Future Success*, Information Systems Engineering and Management 22,
https://doi.org/10.1007/978-3-031-76152-2_11

adjustments cater to each sector's distinct difficulties and possibilities, ensuring governance frameworks meet industrial needs. In conclusion, this study sheds light on smart system governance frameworks and the relationship between engineering and information management. The research provides customizable frameworks that blend technical advances with ethical and transparent governance to provide the groundwork for smart system success.

Keywords Smart systems · Governance frameworks · Transparent governance · Data integrity · Healthcare

1 Introduction

The use of smart technologies in many parts of our lives has grown more commonplace in this fast-developing digital era. Automated industrial processes, linked healthcare systems, smart cities, and transportation networks are just a few examples of the many applications of smart technology that hold the promise of greater sustainability, ease of use, and efficiency. On the other hand, strong regulatory frameworks that deal with the complicated problems and ethical concerns related to the expansion of smart systems are just as important as technology progress in making these advantages a reality. A key function of smart system governance frameworks is to guarantee the ethical and responsible deployment of smart technology by established social norms and values [1]. All sorts of rules, laws, standards, and guidelines are incorporated into these frameworks to control the creation, implementation, and maintenance of smart systems in different fields. Basic to good smart system governance frameworks is the pursuit of a happy medium between encouraging innovation and protecting against hazards and unforeseen effects [2]. While simultaneously encouraging participation from all relevant parties in decision-making and navigating the complex terrain of data privacy, cybersecurity, algorithmic transparency, accountability, and equity, they face several challenges. This in-depth analysis of smart system governance frameworks will examine the fundamentals, principles, and difficulties of controlling the introduction of these game-changing technologies. Our goal is to shed light on the complex nature of smart system governance by analyzing case studies, new best practices, and interdisciplinary viewpoints. Academics, lawmakers, and business leaders will all benefit from a clearer understanding of the subject as a result of this. While ensuring that these technologies are used for the greater good and maintain fundamental ethical principles, proactive and adaptive governance may unleash the potential of smart systems to address pressing societal concerns, encourage innovation, and raise living standards. To successfully navigate the complex terrain of the digital age, it is critical to establish robust governance frameworks that allow society to profit from smart technology appropriately and sustainably.

1.1 Background

A network of physical items connected by software, sensors, and other technologies is called a "smart system". With the help of these components, the system can collect, share, and evaluate data in addition to reacting quickly to human inputs and environmental changes. Artificial Intelligence (AI), Machine Learning (ML), the Internet of Things (IoT), and data analytics are some of the technologies that smart systems use to increase their usefulness, efficiency, and autonomy. They can sense their environment, assimilate information instantly, and use that knowledge to forecast or make judgments. They can frequently operate independently to complete particular activities or goals. Many industries, including smart cities, manufacturing, transportation, healthcare, agriculture, and home automation, use smart systems. In essence, smart systems strive to enhance efficiency, optimize resource utilization, improve user experiences, and introduce new capabilities by integrating advanced technologies and intelligent decision-making mechanisms.

A new way of thinking about and interacting with technology and the physical environment has emerged with the advent and quick development of smart systems. These systems, made possible by state-of-the-art technologies like cloud computing, big data analytics, the Internet of Things (IoT), and Artificial Intelligence (AI), could completely transform numerous industries, such as healthcare, transportation, urban planning, energy management, agriculture, and manufacturing [3]. The three pillars of smart systems—intelligence, automation, and interconnection—form its foundation as shown in Fig. 1. These systems can gather, analyze, and act upon massive amounts of data in real-time by embedding computational capabilities, actuators, and sensors into physical objects and environments. This allows for better decision-making, better utilization of resources, and increased productivity and efficiency in general [4]. As an example, in the setting of smart cities, sensor networks may track things like energy consumption, environmental factors, and traffic flow. This enables smarter resource allocation, more tailored services for inhabitants, and dynamic control of transportation systems. Wearables and other remote monitoring technologies in healthcare are highly useful in monitoring vital signs, early diagnosis of health problems, and the creation of individualized treatment programs. This, in turn, may improve patient outcomes while reducing healthcare expenditures.

But there are a lot of things to think about and problems to solve when smart systems are widely used, especially when it comes to the ethical and governmental aspects. To minimize risks and guarantee that smart systems are implemented in a way that encourages fairness, openness, and responsibility, it is necessary to thoroughly examine concerns including data privacy, cybersecurity vulnerabilities, algorithmic bias, digital divide, and social effect. Given this context, smart system governance frameworks have become an important topic of discussion among academics, legislators, and business executives. These frameworks aim to provide a path for smart technology development, implementation, and operation that is both innovative and respectful of society's values and ethics. Governance frameworks aim to address privacy, security, fairness, and sustainability problems by laying out transparent

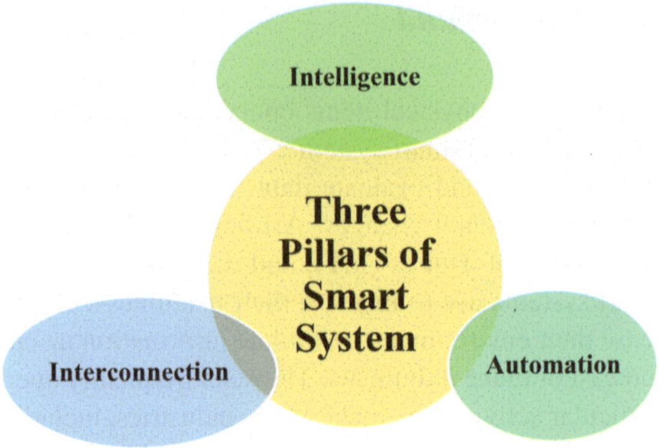

Fig. 1 Three pillars of smart system

laws, regulations, norms, and guidelines; they also want to build confidence among stakeholders and encourage responsible innovation [5]. To benefit society and the public interest, smart systems make it simpler for government agencies, corporations, universities, and community organizations to collaborate. Everyone will examine smart system governance frameworks' core, principles, and challenges, examining their implications in diverse industries and how to adopt and adapt to a continuously changing technological environment.

1.2 Significance of Governance in Smart Systems

Governance frameworks for smart systems, while crucial for their effective and responsible operation, face several significant challenges. These difficulties arise from the complex nature of smart systems, the rapid pace of technological advancement, and the diverse range of stakeholders involved. Below are some key challenges in governing smart systems, particularly concerning engineering and managing information for future success such as Technological Complexity, Data Management and Privacy, Interoperability and Standards Cybersecurity Risks, Regulatory and Legal Challenges, Ethical and Societal Implications, and Public Trust and Engagement. The significance of governance in smart systems cannot be overstated, as it plays a crucial role in shaping the development, deployment, and impact of these transformative technologies. Several key factors highlight the importance of governance in the context of smart systems:

- ***Ethical and Responsible Use***: Smart systems often involve the collection, processing, and utilization of vast amounts of data, raising important ethical considerations regarding privacy, consent, fairness, and accountability [6]. Governance frameworks help ensure that these technologies are developed and deployed in a manner that respects individual rights, promotes transparency, and upholds ethical standards.

- ***Risk Management and Security***: Smart systems are susceptible to various risks, including cyberattacks, data breaches, and system failures, which can have noteworthy significance for individuals, organizations, and society as a whole. Governance frameworks establish measures to mitigate these risks, including robust cybersecurity protocols, data encryption, access controls, and contingency plans to address potential threats and vulnerabilities.
- ***Compliance and Regulation***: Given the complexity and novelty of smart systems, regulatory frameworks are essential to ensure compliance with legal requirements, industry standards, and ethical guidelines. Governance mechanisms help navigate the regulatory landscape, providing clarity on issues such as data protection, intellectual property rights, safety standards, and liability, thereby reducing legal uncertainties and fostering innovation.
- ***Accountability and Transparency***: Governance frameworks promote accountability among shareholders complex in the development, deployment, and operation of smart systems. By establishing clear roles, responsibilities, and mechanisms for oversight, they hold organizations accountable for their actions and decisions, while also ensuring transparency in how data is collected, processed, and used.
- ***Equity and Inclusivity***: Smart systems have the potential to exacerbate existing inequalities and create digital divides if not implemented thoughtfully. Governance frameworks address concerns related to equity and inclusivity, ensuring that the benefits of these technologies are accessible to all members of society and that vulnerable populations are not marginalized or disadvantaged.
- ***Stakeholder Engagement and Collaboration***: Effective governance involves engaging a diverse range of stakeholders, including government agencies, industry partners, academia, civil society organizations, and the public. By fostering collaboration and dialogue among these stakeholders, governance frameworks facilitate the co-design and co-implementation of smart systems, incorporating diverse perspectives.
- ***Long-term Sustainability***: Sustainable development is a key consideration in the deployment of smart systems, as their widespread adoption can have significant environmental, social, and economic implications. Governance frameworks promote sustainability by encouraging the use of renewable energy sources, minimizing resource consumption, reducing waste and pollution, and promoting the circular economy principles.

In essence, governance frameworks serve as the cornerstone for ensuring that smart systems are developed and deployed in a manner that maximizes their benefits while minimizing risks and adverse impacts. By establishing clear principles, standards, and mechanisms for oversight and accountability, governance frameworks provide the necessary foundation for responsible innovation and the ethical use of technology in advancing societal well-being. The study's overarching goal is to get a better understanding of the relevance and consequences of smart system governance frameworks in different fields by delving into their complex environment. In doing so, it hopes to better understand the building blocks of such frameworks, the principles and best practices that support their development and execution, and the threats and difficulties that come with regulating the rollout of smart technology. This research endeavors to fill a gap in our knowledge of smart system governance by using an interdisciplinary approach and analyzing it via the lenses of law, ethics, technology, sociology, and public policy, among others. It will examine healthcare, smart city, and transportation case studies and examples to show how various governance models affect real-world situations. In conclusion, the study aims to provide scholars, industry stakeholders, and policymakers with practical suggestions for improving governance frameworks that promote responsible innovation and preserve society values in the smart system age [7].

2 Literature Review

The governance frameworks for smart systems have been a topic of interest in recent literature. Rottondi et al. [8] discussed social computing applications for smart cities, highlighting the importance of incorporating technology into urban environments. Milchram et al. [9] examined the design options in smart grid projects and their influence on energy justice, underlining the importance of explicit standards for designers and politicians. Darmawan et al. [10] investigated the success elements of mobile-based smart regency services, which are consistent with smart city development concepts. Fiorentino et al. [11] proposed blockchain-based smart contracts as new governance tools for the sharing economy, intending to increase transaction traceability and transparency. Biyik et al. [12] evaluated the literature on smart parking systems, highlighting the need for large-scale vehicle detection technologies in the creation of intelligent solutions.

Taratori et al. [13] explored the emergence of innovation ecosystems in modern cities, as well as the acceptance of configurational approaches to smart governance models. Hentea [14] discussed the problems and privacy issues posed by the smart grid, highlighting the importance of legal frameworks and proper protection. Hine et al. (2022) discussed ethical concerns in the design and implementation of home-based smart care for dementia, emphasizing the significance of ethical assessment and governance in healthcare [15]. Lane (2023) highlights the role of private AI developers in mitigating long-term hazards to human rights in smart cities, highlighting the need for ethical concerns in AI system development [16]. Overall, the literature review on governance frameworks for smart systems covers a range of topics including social computing applications, energy justice in smart grids, success factors for smart services, blockchain-based governance tools, smart parking systems, innovation ecosystems, ethical considerations in healthcare, and the responsibilities of private AI developers in smart city technologies. These studies collectively contribute to the understanding of governance frameworks for smart systems and highlight the importance of ethical, transparent, and inclusive approaches in their development and implementation.

2.1 *Evolution of Smart Systems*

Smart systems represent a milestone in the confluence of technology, connection, and intelligence, changing how individuals interact with the environment. Smart systems started as single-purpose gadgets with limited connection and intelligence. However, advancements in key technologies have fueled the evolution of smart systems into highly interconnected, intelligent ecosystems capable of sensing, analyzing, and responding to diverse inputs in real-time [17]. The evolution of smart systems (Fig. 2) can be traced through several key stages:

- **Emergence of Embedded Systems**: The early stages of smart systems were marked by the development of embedded systems, where computing capabilities were integrated into physical objects to perform specific tasks. Examples include early industrial automation systems and simple consumer electronics with basic programmable features.
- **Connectivity and Interoperability**: As internet connectivity became more widespread, smart systems evolved to incorporate networked capabilities, and enable data exchange via communication between devices and systems. This phase saw the emergence of the Internet of Things (IoT), where sensors and actuators embedded in various objects could communicate with each other and with central servers or cloud platforms.
- **Data-driven Intelligence**: With the proliferation of connected devices generating vast amounts of data, the focus shifted towards leveraging advanced analytics for extracting actionable insights from this data. This phase saw the rise of data-driven intelligence, where smart systems could analyze complex datasets in real-time to optimize processes, predict outcomes, and enable more informed decision-making.
- **Autonomy and Adaptability**: Building upon data-driven intelligence, the latest phase of smart system evolution involves the development of autonomous and

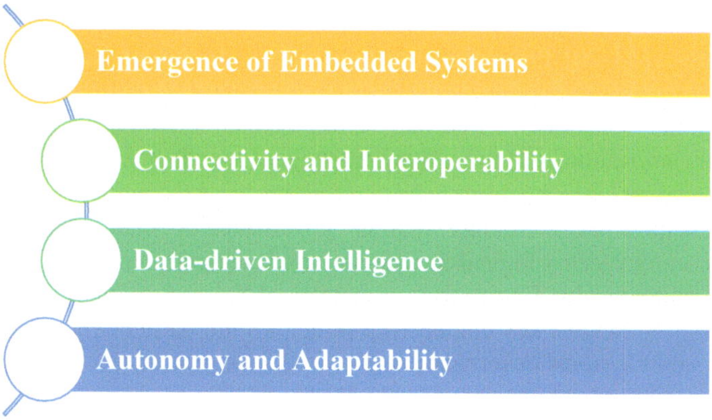

Fig. 2 Evolution of smart systems

adaptive systems that can learn from experience, optimize their behavior, and adapt to changing environments autonomously. This includes applications such as autonomous vehicles, smart grids, and intelligent personal assistants that leverage AI and machine learning to continuously improve performance and user experience [18–20].

Smart technologies have become widespread in smart cities, healthcare, transportation, energy, agriculture, and manufacturing, changing how humans tackle issues and possibilities in each area. Smart systems will continue to evolve due to advances in 5G connectivity, edge computing, robotics, and quantum computing, which will blur the lines between the physical and digital worlds and create new opportunities for innovation and disruption [21–27].

3 Engineering Principles in Smart Systems Governance

The creation, implementation, and operation of these complex technological ecosystems are guided by engineering principles, which are essential in the governance of smart systems. The need for engineers to collaborate with legislators, attorneys, ethicists, and other stakeholders to guarantee all-encompassing governance solutions highlights the importance of interdisciplinary cooperation. To promote trust and understanding among stakeholders, organizations advocate for transparent decision-making and accountability processes. Privacy by design emphasizes integrating encryption, data minimization, and user consent into smart technologies. To safeguard smart systems against cyber-attacks and make them resilient to disturbances, resilience engineering concepts stress the need to include strong security measures in smart system designs from the start. When designing and deploying smart systems, engineers should keep ethical concerns in mind to ensure the safety of people and communities, prevent damage, and advance justice and fairness. Stakeholders can evaluate and modify governance frameworks in response to changing societal demands and real-world performance data via continuous review and improvement processes [18–20]. Principles of sustainability and environmental responsibility motivate the creation of eco-friendly smart systems that support the use of renewable energy, the reduction of waste, and the implementation of the circular economy. Responsible innovation, risk mitigation, and societal benefit maximization may all be achieved by smart systems governance that follows these technical principles. Engineering principles play a crucial role in the governance of smart systems, ensuring that these complex technological ecosystems are developed, deployed, and operated effectively and ethically. Engineering principles of Smart Systems Governance are shown in Fig. 3. Several key engineering principles are particularly relevant in the context of smart systems governance:

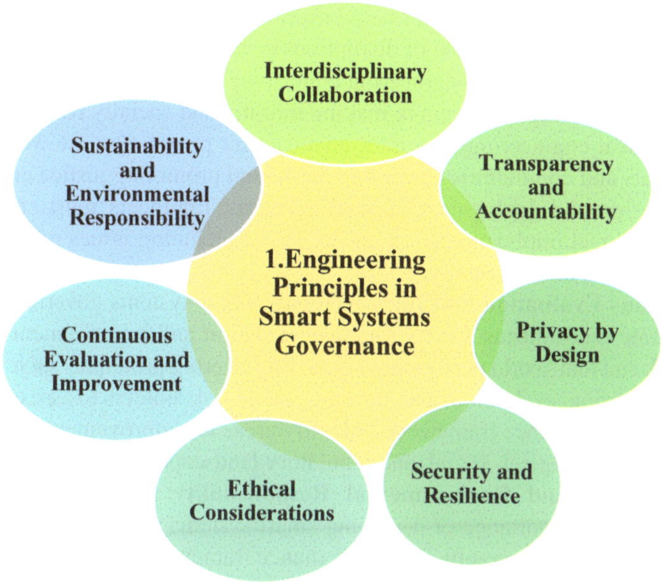

Fig. 3 Engineering principles of smart systems governance

- **Interdisciplinary Collaboration**: Smart systems governance often requires collaboration among engineers, policymakers, legal experts, ethicists, and other stakeholders. By fostering interdisciplinary collaboration, engineers can ensure that governance frameworks account for diverse perspectives and considerations, leading to more comprehensive and effective governance solutions.
- **Transparency and Accountability**: Engineering principles emphasize the importance of transparency and accountability in the design and operation of smart systems. Governance frameworks should include mechanisms for transparent decision-making, open communication, and clear accountability structures to ensure that stakeholders understand how smart systems operate and who is responsible for their outcomes.
- **Privacy by Design**: Privacy is a critical consideration in smart systems governance, particularly given the vast amounts of data collected and processed by these technologies. Engineering principles such as privacy by design advocate for integrating privacy protections into the design of smart systems from the outset, rather than as an afterthought. This involves minimizing data collection, implementing strong encryption and access controls, and providing users with meaningful consent mechanisms.
- **Security and Resilience**: To guard against dangers like data breaches and cyberattacks, smart systems need to be built with strong security features. The significance of incorporating security elements, such as encryption, authentication, and intrusion detection measures, directly into the architecture of smart systems is emphasized by engineering principles such as security by design. Additionally,

resilience engineering principles help ensure that smart systems can continue to operate effectively in the face of disruptions or failures.

- **Ethical Considerations**: Engineering ethics play a central role in smart systems governance, guiding engineers in making morally and socially responsible decisions. Ethical engineering principles advocate for prioritizing the well-being of individuals and communities, avoiding harm, and promoting justice and fairness in the design and deployment of smart systems. Engineers must consider the potential ethical implications of their decisions, including issues related to bias, discrimination, and autonomy.
- **Continuous Evaluation and Improvement**: Smart systems governance should incorporate mechanisms for continuous evaluation and improvement based on feedback and real-world performance data. Engineering principles such as continuous monitoring, evaluation, and iteration enable stakeholders to assess the effectiveness of governance frameworks, identify areas for improvement, and adapt to evolving technological, social, and regulatory landscapes.
- **Sustainability and Environmental Responsibility**: Engineering principles emphasize the importance of designing smart systems that are environmentally sustainable and resource-efficient. Governance frameworks should encourage the use of renewable energy sources, minimize waste and pollution, and promote circular economy principles to ensure that smart systems contribute to long-term environmental sustainability.

Stakeholders may make sure that these technologies are developed and implemented in a way that optimizes their advantages while avoiding risks and negative effects on people, communities, and the environment by adhering to these engineering principles in smart systems governance. There are two specific engineering approaches commonly employed in governance frameworks for smart systems:

- **Secure Systems Engineering**: Secure systems engineering involves integrating security measures into the design, development, and operation of smart systems to protect against unauthorized access, data breaches, and cyber threats. This approach encompasses implementing encryption protocols, access control mechanisms, intrusion detection systems, and secure software development practices to safeguard sensitive data and ensure the integrity and confidentiality of system operations.
- **Real-Time Systems Engineering**: Real-time systems engineering focuses on designing smart systems capable of processing and responding to input within specified time constraints. Applications like driverless cars, industrial automation, and healthcare monitoring systems that demand quick data processing and decision-making must use this strategy. Optimizing system design, hardware parts, and software algorithms to reduce latency and guarantee dependable performance in dynamic situations is the goal of real-time systems engineering.

3.1 Systems Engineering for Smart Systems

Systems engineering offers an organized method for creating and overseeing intricate technical ecosystems, which is vital to the creation, implementation, and functioning of smart systems. Systems engineering concepts are crucial in the context of smart systems, which incorporate several interrelated components and stakeholders, to guarantee that these systems operate efficiently, dependably, and by user requirements and societal objectives.

Requirements management and analysis are essential components of systems engineering for intelligent systems. This entails determining and recording the goals and demands of stakeholders, converting them into technical specifications, and overseeing modifications made to the system over its entire existence [9]. Requirements for smart systems may include functionalities, performance indicators, standards for interoperability, requirements for data security and privacy, and adherence to legal requirements.

System architecture design is a crucial component of systems engineering for intelligent systems. This entails specifying the system's architecture, including its parts, interfaces, and interactions, to maximize performance, scalability, and reliability while satisfying the requirements. The design of system architecture for intelligent systems often takes into account factors like data interchange formats, distributed computing, edge computing, and interoperability protocols to facilitate smooth integration and communication across various systems and components. The creation and integration of hardware and software components inside smart systems is also included in the field of systems engineering. This covers hardware selection, configuration, and validation in addition to software design, development, testing, and integration. Systems engineers must make sure that hardware and software components interact smoothly with smart systems, which often depend on cutting-edge technology like artificial intelligence, machine learning, and Internet of Things sensors, to achieve the required functionality and performance. Systems engineering for smart systems also includes thorough testing, validation, and verification to guarantee that the system satisfies specifications and operates as intended. This might include a range of testing methods, including simulation and modeling to evaluate how the system will behave in various situations and settings, as well as unit, integration, system, and user acceptability testing.

Systems engineering for smart systems also takes into account how the system will operate, maintain itself, and change over time. To guarantee the system's continued performance, availability, and dependability, this entails creating maintenance schedules, monitoring and management tools, and operating protocols. To meet changing user demands, technical breakthroughs, and regulatory requirements, it also includes planning for system expansions, upgrades, and improvements. All things considered, systems engineering offers an organized and comprehensive method for creating, constructing, and overseeing intelligent systems, guaranteeing that they satisfy user requirements, adhere to legal requirements, and make a meaningful and long-lasting contribution to the welfare of society. Stakeholders may handle the complexity and unpredictability involved in the development and deployment of smart systems by using systems engineering concepts, which will eventually result in more successful solutions and larger benefits for society.

3.2 *Integration of Artificial Intelligence and IoT*

Figure 4 illustrates how Artificial Intelligence (AI) and the Internet of Things (IoT) are merging into one powerful technological force that can change a wide range of industries and fields. This integration involves embedding AI algorithms and capabilities inside IoT devices and systems to allow them to collect, analyze, and act upon data in real-time to make intelligent decisions and automate activities. The integration of AI and IoT is characterized by several important factors.

- ***Data Acquisition and Sensing***: IoT devices serve as the foundation for collecting data from the physical world, including sensors, actuators, cameras, and other monitoring devices. These devices capture a wide range of data, such as environmental conditions, equipment performance, and user behavior, generating massive streams of raw data that can be analyzed and processed by AI algorithms.
- ***Data Analysis***: Large volumes of data produced by Internet of Things devices are analyzed by AI algorithms to glean insights from the data. This entails using methods like deep learning, machine learning, and natural language processing to find trends, anomalies, correlations, and patterns in the data. IoT systems may

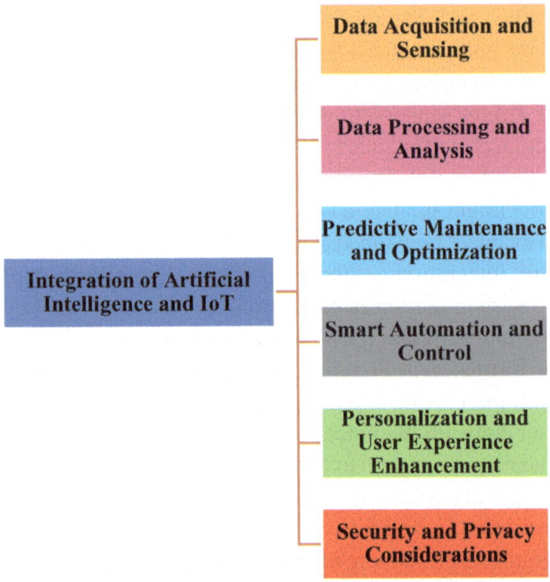

Fig. 4 Integration of artificial intelligence and IoT

obtain useful insights and make wise decisions instantly by utilizing AI-driven analytics.

- ***Predictive Maintenance and Optimization***: Predictive maintenance, in which AI algorithms examine sensor data to find early indicators of equipment failure or degradation, is one of the principal uses of AI and IoT integration. Organizations may maximize asset performance, limit downtime, and save maintenance costs by anticipating maintenance needs before they arise. Similar to this, AI-driven optimization algorithms may maximize productivity and resource use by dynamically modifying system setups and parameters based on real-time data.
- ***Smart Automation and Control***: IoT devices with AI capabilities allow for intelligent automation and management of physical systems and processes. Organizations can automate repetitive activities, manage energy usage, and boost operational efficiency by incorporating AI algorithms into IoT devices and edge computing platforms. For instance, industrial robots employ AI-driven vision systems to carry out difficult jobs with accuracy and precision, while smart thermostats use AI algorithms to learn user preferences and adjust temperature settings automatically.
- ***Personalization and User Experience Enhancement***: AI and IoT integration enables personalized and context-aware experiences for users across various domains. By analyzing user data and preferences collected from IoT devices, AI algorithms can deliver personalized recommendations, content, and services tailored to individual needs and preferences. This could include personalized healthcare interventions based on real-time health data, customized product recommendations in retail environments, or adaptive lighting and temperature settings in smart homes.
- ***Security and Privacy Considerations***: The integration of AI and IoT raises important security and privacy considerations, given the sensitivity of the data involved and the potential impact of security breaches. AI-driven security solutions can help detect and mitigate cybersecurity threats in IoT networks while privacy-preserving AI techniques can anonymize and encrypt sensitive data to protect user privacy.

Overall, the integration of AI and IoT holds immense potential to transform industries, enhance productivity, and improve quality of life. By harnessing the power of AI-driven analytics and intelligent automation, IoT systems can unlock new capabilities and drive innovation across a wide range of applications, from smart cities and healthcare to manufacturing and agriculture. However, to realize the full benefits of this integration, organizations must carefully consider the design, implementation, and governance of AI-powered IoT systems, ensuring that they are secure, ethical, and aligned with user needs and societal values.

3.3 Cybersecurity Considerations

The complex interconnectedness and data-rich nature of AI and IoT integration make cybersecurity issues critical. The security of the data that AI-enabled Internet of Things devices create and send is crucial to this. In addition to robust access restrictions and authentication procedures to prevent unwanted access, strong encryption methods must be used to protect data both in transit and at rest. Furthermore, to confirm the legitimacy of IoT devices and control their network access rights, device authentication and authorization protocols are crucial. Encrypting data transfer and preventing efforts at manipulation and eavesdropping are made possible by secure communication protocols like Transport Layer Security (TLS) or Message Queuing Telemetry Transport (MQTT). To strengthen device security and fix known vulnerabilities, regular firmware and software upgrades are essential. Moreover, edge computing—which is often used in AI-enabled IoT systems—needs strong security protocols, such as intrusion detection systems and anomaly detection algorithms, to strengthen local processing capacities. Machine learning algorithms are widely used in AI-driven Internet of Things applications. To protect them against adversarial attacks and data poisoning, input validation and anomaly detection are necessary. Ensuring privacy is of utmost importance, which calls for the use of privacy-preserving AI methods such as differential privacy to anonymize data without compromising its usefulness. Preventing supply chain threats also requires guaranteeing the security of the whole supply chain, from the production of devices to their deployment. In conclusion, a thorough incident response strategy is essential for allowing quick identification, action, and recovery from security breaches to save downtime and safeguard system integrity. Organizations may create resilient AI-enabled IoT ecosystems that efficiently use data-driven insights while reducing the associated risks of cyber-attacks by taking a diversified approach to address these cybersecurity factors.

3.4 Scalability and Interoperability

Scalability and interoperability are paramount considerations in the integration of Artificial Intelligence (AI) and the Internet of Things (IoT), facilitating the seamless expansion and integration of interconnected systems. Scalability ensures that AI-enabled IoT ecosystems can accommodate growing data volumes, user bases, and device deployments without sacrificing performance or efficiency. By employing modular design principles and scalable infrastructure architectures, organizations can dynamically scale their AI-enabled IoT systems to meet evolving demands and accommodate new use cases. Interoperability, on the other hand, enables disparate IoT devices, platforms, and technologies to communicate and exchange data effectively, promoting seamless integration and collaboration across heterogeneous environments. Standardization efforts, such as the adoption of common communication protocols and data formats, play a crucial role in enhancing interoperability and facilitating plug-and-play compatibility between different IoT devices and systems. Moreover, open Application Programming Interfaces (APIs) facilitate seamless integration with third-party applications and services, enabling organizations to leverage the broader ecosystem of AI and IoT solutions. Together, scalability and interoperability empower organizations to build agile and adaptive AI-enabled IoT ecosystems that can evolve and expand over time, unlocking new opportunities for innovation and collaboration while maximizing the value of interconnected systems.

4 Information Management Strategies

AI-enabled IoT systems need sophisticated frameworks to handle the vast amounts of data they create and analyze. These frameworks are called information management techniques. These tactics include a range of methods designed to collect, store, handle, process, evaluate, and share data while maintaining security and legal compliance. Prioritizing data integration and acquisition at first entails building strong pipelines for gathering data from many sources and integrating it efficiently. The focus then shifts to data management and storage, where scalable methods like distributed databases and cloud-based storage are used in conjunction with data management techniques to maximize effectiveness. Real-time processing approaches effectively manage streaming data, while processing and analysis take center stage, using AI and machine learning algorithms to extract actionable insights from the acquired data. Throughout, security and privacy remain top priorities, with data protection guaranteed by encryption, access restrictions, and adherence to laws like General Data Protection Regulation (GDPR) and Health Insurance Portability and Accountability Act (HIPAA). Strong data governance structures are also put in place to guarantee compliance and integrity, and continued efficacy and dependability are guaranteed by techniques for continual monitoring and development. Through the implementation of these techniques, enterprises may fully use the promise of AI-powered IoT devices, all the while protecting their data assets and adhering to legal requirements. Information management strategies are essential components of effectively handling the vast volumes of data generated by AI-enabled IoT systems. These strategies encompass various practices and methodologies aimed at acquiring, storing, processing, analyzing, and disseminating data in a manner that is secure, efficient, and conducive to actionable insights. Key components of information management strategies in the context of AI and IoT integration include:

- **Data Acquisition and Integration**: This involves collecting data from diverse sources, such as sensors, devices, and external systems, and integrating it into a unified data ecosystem. Strategies for data acquisition and integration include establishing data pipelines, leveraging APIs and middleware for interoperability, and employing protocols for standardized data exchange.

- **Data Storage and Management**: Given the sheer volume and variety of data generated by AI-enabled IoT systems, effective data storage and management strategies are crucial. This includes selecting appropriate storage solutions such as cloud-based storage, distributed databases, or data lakes, and implementing data management practices such as data deduplication, compression, and indexing to optimize storage efficiency and accessibility.
- **Data Processing and Analysis**: Data processing and analysis are central to extracting actionable insights from IoT data. Strategies in this area encompass employing AI and machine learning algorithms for predictive analytics, anomaly detection, and pattern recognition, as well as utilizing techniques such as stream processing and real-time analytics to handle data in motion.
- **Data Security and Privacy**: It is critical to prevent unauthorized access, alteration, and disclosure of sensitive data. Implementing encryption, access limits, and data anonymization techniques are some strategies for guaranteeing data security and privacy. Other strategies include complying with legal regulations like GDPR and HIPAA.
- **Data Governance and Compliance**: Establishing robust data governance frameworks is essential for ensuring data integrity, quality, and compliance with relevant regulations and standards. This involves defining policies and procedures for data management, assigning roles and responsibilities, and implementing mechanisms for data lineage, auditing, and compliance monitoring.
- **Data Visualization and Reporting**: Communicating insights derived from IoT data effectively is essential for facilitating decision-making and driving action. Strategies for data visualization and reporting include using dashboards, charts, and interactive visualizations to present complex data in a clear and understandable format, as well as employing reporting tools and systems for sharing insights and Key Performance Indicators (KPIs) across the organization.
- **Continuous Monitoring and Improvement**: Continuous monitoring of data quality, performance, and security is necessary to ensure the ongoing effectiveness and reliability of AI-enabled IoT systems. Strategies for continuous monitoring and improvement include implementing monitoring tools and dashboards for real-time visibility into system health, as well as establishing feedback loops and iterative improvement processes to refine data management strategies and algorithms over time.

By implementing these information management strategies, organizations can effectively leverage the power of AI and IoT technologies to derive actionable insights, optimize operations, and drive innovation while ensuring the security, privacy, and integrity of their data assets. The following section describes how iteration can be incorporated into the design of governance frameworks for smart systems, along with a brief example:

- **Iterative Design Process**: The governance frameworks can adopt an iterative design process, where they are continuously refined and improved based on feedback, new insights, and evolving requirements. This iterative approach allows stakeholders to gradually enhance the effectiveness, adaptability, and relevance of the frameworks over time. Example: A governance framework for smart transportation systems undergoes iterative development through multiple phases. In the initial phase, stakeholders collaborate to define system requirements, identify potential risks, and propose governance mechanisms. Subsequently, the framework is piloted in a small-scale transportation project, allowing stakeholders to evaluate its effectiveness, identify gaps, and gather feedback from users and experts. Based on the lessons learned from the pilot implementation, the framework is iterated upon to address identified issues, incorporate new insights, and adapt to changing regulatory requirements or technological advancements. This iterative process continues through subsequent deployments, with each iteration enhancing the maturity and resilience of the governance framework.

- **Feedback Loops**: The governance frameworks can incorporate feedback loops to gather input from stakeholders, monitor system performance, and identify opportunities for improvement. By establishing feedback mechanisms, such as surveys, reviews, and performance metrics, stakeholders can continuously assess the effectiveness and impact of the frameworks and make informed decisions about adjustments or enhancements. Example: A governance framework for smart energy grids includes feedback loops to solicit input from utility providers, regulators, and consumers. Utility providers regularly collect data on system reliability, efficiency, and customer satisfaction, which is used to evaluate the effectiveness of the governance framework. Additionally, regulators conduct periodic reviews and assessments of the framework's compliance with regulatory requirements and industry standards. Based on the feedback received from stakeholders and performance metrics, adjustments are made to the governance framework to address emerging challenges, improve system performance, and enhance stakeholder engagement.

By incorporating iterative design processes and feedback loops into the governance frameworks, stakeholders can ensure that they remain responsive to changing needs, emerging technologies, and evolving regulatory landscapes, thereby enhancing their effectiveness and relevance in governing smart systems.

4.1 Data Governance and Integrity

Data governance and integrity are fundamental aspects of information management within AI-enabled IoT systems, ensuring that data is reliable, accurate, and secure throughout its lifecycle. Data governance encompasses the policies, procedures, and controls put in place to manage data effectively, while data integrity refers to the assurance that data is accurate, consistent, and trustworthy. Together, they form the foundation for building trust in the data and ensuring its usability for decision-making and analytics. At its core, data governance involves defining roles, responsibilities, and processes for managing data assets within an organization. This includes establishing data ownership, data stewardship, and data quality standards to ensure that data is properly managed, protected, and utilized. Data governance frameworks also encompass policies for data access, sharing, and usage, ensuring that data is accessed only by authorized users for legitimate purposes. Data integrity, on the other hand, focuses on maintaining the accuracy and consistency of data throughout its lifecycle. This involves implementing controls and measures to prevent data corruption, unauthorized modifications, and data quality issues. Techniques such as data validation, checksums, and encryption are used to verify the integrity of data at various stages, from acquisition and storage to processing and analysis. In the context of AI-enabled IoT systems, data governance, and integrity are particularly critical due to the diverse sources and types of data involved, as well as the potential for data to be collected, processed, and shared across multiple devices and platforms. Robust data governance frameworks help ensure that data is managed consistently and in compliance with regulatory requirements, while data integrity measures safeguard against data tampering, corruption, and unauthorized access. Key components of effective data governance and integrity (Fig. 5) in AI-enabled IoT systems include:

The following section shows how the principles of data integrity, privacy, and accessibility can be operationalized within the proposed governance frameworks for smart systems:

- **Data Integrity**: Within the governance frameworks, data integrity can be operationalized through the implementation of encryption mechanisms, data validation techniques, and integrity checks. This ensures that data remains accurate, consistent, and trustworthy throughout its lifecycle. Additionally, regular audits and monitoring processes can be established to detect and mitigate any instances of data corruption or tampering.
- **Privacy**: To operationalize privacy principles, the governance frameworks can incorporate privacy-enhancing technologies such as anonymization,

Fig. 5 Data governance and integrity

pseudonymization, and differential privacy. These techniques help protect individuals' privacy by minimizing the risk of re-identification and unauthorized access to sensitive personal information. Furthermore, privacy impact assessments and privacy-by-design approaches can be integrated into the system development process to identify and mitigate privacy risks proactively.

- **Accessibility**: Within the governance frameworks, accessibility principles can be operationalized by ensuring that smart systems are designed and implemented in compliance with accessibility standards such as the Web Content Accessibility Guidelines (WCAG). This involves making user interfaces, communication protocols, and interaction modalities inclusive and accessible to individuals with diverse abilities and needs. Additionally, providing alternative formats, assistive technologies, and user support services can enhance the accessibility of smart systems for all users.

By operationalizing these principles within the proposed governance frameworks, stakeholders can ensure that smart systems prioritize data integrity, privacy protection, and accessibility for all users, thereby fostering trust, transparency, and accountability in the deployment and operation of smart systems.

4.2 Transparency and Accountability Mechanisms

Transparency and accountability mechanisms are essential components of governance within AI-enabled IoT systems, ensuring that decision-making processes are open, fair, and accountable. Transparency involves making information about the design, operation, and outcomes of AI and IoT systems accessible to stakeholders, while accountability ensures that individuals and organizations are held responsible for their actions and decisions. Together, they promote trust, fairness, and ethical conduct within AI-enabled IoT ecosystems. Transparency mechanisms in AI-enabled IoT systems include providing clear documentation about system architecture, data sources, algorithms, and decision-making processes. This enables stakeholders to understand how AI and IoT technologies are being used, what data is being collected and processed, and how decisions are made based on that data. Additionally, transparency may involve providing explanations or justifications for AI-driven decisions, particularly in high-stakes applications such as healthcare or autonomous vehicles.

Accountability mechanisms, on the other hand, hold individuals and organizations responsible for their actions and decisions within AI-enabled IoT systems. This may involve assigning roles and responsibilities for data governance, security, and compliance, as well as establishing mechanisms for oversight, review, and enforcement of policies and regulations. Accountability mechanisms also include mechanisms for recourse and redress in the event of errors, biases, or malfunctions in AI-driven decision-making processes. Key components of transparency and accountability mechanisms in AI-enabled IoT systems include:

- ***Governance Structures***: Establishing clear governance structures and accountability frameworks to oversee the development, deployment, and operation of AI-enabled IoT systems. This may involve creating cross-functional teams, advisory boards, or steering committees responsible for decision-making and oversight.
- ***Transparency Reports***: Publishing regular reports or documentation about AI and IoT systems, including information about data collection practices, algorithmic decision-making processes, and performance metrics. Transparency reports help stakeholders understand how AI and IoT technologies are being used and ensure that decision-making processes are open and transparent.
- ***Ethical Guidelines and Codes of Conduct***: Developing and adhering to ethical guidelines and codes of conduct that govern the use of AI and IoT technologies. This may include principles such as fairness, transparency, accountability, and privacy by design, which guide decision-making and behavior within AI-enabled IoT ecosystems.
- ***Auditing and Certification***: Conduct regular audits and assessments of AI-enabled IoT systems to evaluate their compliance with ethical standards, regulatory requirements, and best practices. Certification programs and independent audits help validate the transparency and accountability of AI and IoT technologies and build trust among stakeholders.
- ***User Empowerment and Consent***: Empowering users to understand and control how their data is used within AI-enabled IoT systems. This may involve providing clear information about data collection practices, obtaining informed consent for data processing activities, and giving users the ability to opt-out or adjust their privacy settings.

Overall, transparency and accountability mechanisms are essential for ensuring the responsible and ethical use of AI and IoT technologies. By promoting openness, fairness, and accountability, these mechanisms help build trust among stakeholders, mitigate risks, and foster the responsible adoption and deployment of AI-enabled IoT systems for the benefit of society.

5 Stakeholder Engagement

To ensure that decision-making processes take into account the different interests, viewpoints, and demands of stakeholders, stakeholder engagement is a vital component of governance within AI-enabled IoT systems. Some examples of stakeholders in AI-enabled IoT systems include developers, data subjects, legislators, regulators, advocacy organizations, and end-users. Stakeholders should be actively involved in AI and IoT project lifecycle activities, from initial ideation to final assessment, for effective engagement to occur. Asking stakeholders for their thoughts and opinions is an important part of engaging them in the decision-making process. To learn about the goals, worries, and preferences of stakeholders about AI-enabled

IoT systems, it may be necessary to undertake surveys, focus groups, or consultations. To make sure that AI-enabled IoT systems are in line with society's values and objectives, organizations should include stakeholders in the decision-making process. This way, everyone's requirements and expectations may be considered. Fostering openness and dialogue between stakeholders and decision-makers is another critical component of stakeholder engagement. One aspect of this is making sure that information on the goals, features, and possible effects of AI-enabled IoT devices is easily available and understandable. Stakeholders' confidence and credibility may be enhanced by open and honest communication, which in turn encourages a cooperative and fruitful discussion about AI and IoT. Stakeholder engagement also includes encouraging stakeholders to work together in partnerships to tackle shared problems and seize shared opportunities. One approach may be to establish advisory boards, working groups, or multi-stakeholder coalitions to address governance challenges around AI and the Internet of Things. These entities could help with information sharing, capacity development, and collective action. Organizations may improve the quality and longevity of their solutions by working together with stakeholders to pool their knowledge, experience, and resources. In addition, stakeholders can hold decision-makers responsible for their deeds and choices since stakeholder involvement includes systems for feedback and accountability. For AI-enabled IoT systems to work, it's important to provide avenues for stakeholders to voice their concerns, offer feedback, and report problems. Then, be sure to resolve and handle these issues in a transparent and timely way. The governance of AI-enabled IoT systems must prioritize stakeholder involvement if it is to foster inclusion, transparency, and accountability. Organizations can create trust, encourage cooperation, and make sure that AI and IoT technologies are created and used in a way that represents societal values and interests when they include stakeholders in decision-making processes. To foster collaboration among engineers, legislators, industry leaders, and the public, the governance process can incorporate various methods and platforms for stakeholder engagement. This ensures that diverse perspectives are considered, and decisions are made transparently and inclusively. Methods for stakeholder engagement may include:

- **Workshops and Conferences**: Organizing workshops, conferences, and seminars where stakeholders can come together to discuss key issues, share insights, and collaborate on developing governance frameworks. These events provide opportunities for networking, knowledge exchange, and consensus-building among stakeholders from different sectors and backgrounds.
- **Stakeholder Committees and Working Groups**: Establishing stakeholder committees or working groups composed of representatives from engineering organizations, government agencies, industry associations, and community groups. These committees provide a forum for ongoing dialogue, problem-solving, and decision-making on governance issues related to smart systems.
- **Public Consultations and Surveys**: Conducting public consultations and surveys to gather input, feedback, and preferences from members of the public on governance frameworks and policies. This ensures that the interests and concerns of

citizens are considered in the governance process, promoting transparency and accountability.

- **Online Platforms and Social Media**: utilizing social media and internet platforms to promote communication and interaction with stakeholders. This entails setting up specific websites, discussion boards, and social media accounts where interested parties may obtain data, take part in debates, and offer input on governance projects.
- **Partnerships and Collaborative Projects**: Forming partnerships and collaborative projects between engineering organizations, government agencies, academia, industry, and non-profit organizations. These partnerships enable stakeholders to pool resources, share expertise, and work together on joint initiatives to address common challenges and achieve shared goals.

By utilizing these methods and platforms for stakeholder engagement, the governance process can leverage the collective wisdom, expertise, and perspectives of diverse stakeholders to develop inclusive, effective, and sustainable governance frameworks for smart systems.

5.1 Role of Engineers in Governance

The role of engineers in governance within AI-enabled IoT systems is multifaceted, encompassing various responsibilities and contributions throughout the lifecycle of these systems. Engineers play a crucial role in designing, developing, implementing, and maintaining AI-enabled IoT solutions, ensuring that they are technically sound, reliable, and aligned with ethical and regulatory standards. Some key aspects of the role of engineers in governance include:

- **System Design and Architecture**: Engineers are responsible for designing the architecture and components of AI-enabled IoT systems, including hardware devices, software applications, and communication protocols. They must consider factors such as scalability, interoperability, and security when designing these systems to ensure that they can meet the needs of stakeholders and operate effectively within the existing regulatory framework.
- **Ethical Considerations**: Engineers play a vital role in integrating ethical considerations into the design and development of AI-enabled IoT systems. This involves identifying potential ethical issues, such as bias, privacy concerns, and algorithmic transparency, and implementing mechanisms to mitigate these risks. Engineers must adhere to ethical guidelines and principles, such as fairness, transparency, and accountability, to ensure that AI-enabled IoT systems are developed and deployed responsibly.
- **Regulatory Compliance**: Engineers are responsible for ensuring that AI-enabled IoT systems comply with relevant laws, regulations, and standards governing their use. This includes understanding and adhering to data protection regulations (e.g., GDPR), industry-specific standards (e.g., ISO 27001 for information

security), and ethical guidelines (e.g., IEEE Ethically Aligned Design). Engineers must design AI-enabled IoT systems with compliance in mind, incorporating features such as data encryption, access controls, and audit trails to meet regulatory requirements.

- **Risk Management**: Engineers play a key role in identifying and managing risks associated with AI-enabled IoT systems. This includes conducting risk assessments to identify potential hazards and vulnerabilities, implementing safeguards to mitigate these risks, and developing contingency plans to address potential failures or emergencies. Engineers must adopt a proactive approach to risk management, continuously monitoring and updating AI-enabled IoT systems to adapt to evolving threats and regulatory requirements.
- **Collaboration and Communication**: Engineers must collaborate closely with other stakeholders, including policymakers, legal experts, ethicists, and end-users, to ensure that AI-enabled IoT systems are developed and deployed in a manner that meets the needs and expectations of all stakeholders. Effective communication is essential for conveying technical concepts and considerations to non-technical stakeholders and fostering a shared understanding of the risks and benefits of AI-enabled IoT systems.
- **Continuous Improvement**: Engineers play a crucial role in driving continuous improvement and innovation in AI-enabled IoT systems. This involves staying abreast of emerging technologies, best practices, and industry trends, and incorporating new developments into the design and development process. Engineers must also collect and analyze feedback from stakeholders and end-users to identify areas for improvement and refinement in AI-enabled IoT systems.

Overall, the role of engineers in governance within AI-enabled IoT systems is pivotal for ensuring that these systems are developed and deployed responsibly, ethically, and in compliance with regulatory requirements. By integrating ethical considerations, adhering to regulatory standards, managing risks effectively, and fostering collaboration with stakeholders, engineers can contribute to the development of AI-enabled IoT systems that enhance productivity, improve quality of life, and promote societal well-being.

5.2 Policymakers and Regulatory Authorities

Policymakers and regulatory authorities play a crucial role in shaping the governance of AI-enabled IoT systems, establishing the legal and regulatory frameworks that govern their development, deployment, and use. As these technologies continue to advance and become more integrated into various aspects of society, policymakers and regulators face the challenge of balancing innovation with the protection of public interests, including privacy, security, fairness, and ethical considerations. Here are some key roles and responsibilities of policymakers and regulatory authorities in governing AI-enabled IoT systems:

- **Policy Development**: Policymakers are responsible for developing laws, regulations, and guidelines that govern the use of AI-enabled IoT systems. This includes identifying key policy objectives, such as promoting innovation, protecting consumer rights, and ensuring public safety, and translating them into actionable policies that provide clear guidance to stakeholders.
- **Regulatory Oversight**: Regulatory authorities are tasked with overseeing compliance with relevant laws and regulations governing AI-enabled IoT systems. This involves establishing regulatory frameworks, issuing permits and licenses, conducting inspections and audits, and enforcing penalties for non-compliance. Regulatory oversight helps ensure that AI-enabled IoT systems operate safely, securely, and ethically.
- **Risk Assessment and Management**: Policymakers and regulators conduct risk assessments to identify potential hazards and vulnerabilities associated with AI-enabled IoT systems. This includes evaluating risks related to data privacy, security breaches, algorithmic bias, and safety concerns. Based on these assessments, policymakers develop risk management strategies and regulatory requirements to mitigate identified risks and protect public interests.
- **Ethical Guidelines and Standards**: Policymakers and regulatory authorities establish ethical guidelines and standards to govern the development and use of AI-enabled IoT systems. This includes principles such as fairness, transparency, accountability, and human oversight, which guide ethical decision-making and behavior within AI-enabled IoT ecosystems. Ethical guidelines help ensure that AI-enabled IoT systems are developed and deployed in a manner that respects human rights and values.
- **Data Protection and Privacy**: Policymakers and regulators enact laws and regulations to protect the privacy and security of data collected and processed by AI-enabled IoT systems. This includes regulations such as GDPR in Europe and CCPA in California, which impose requirements for data minimization, consent, transparency, and accountability. Data protection laws help ensure that an individual's personal data is handled responsibly and ethically within AI-enabled IoT ecosystems.
- **International Collaboration**: Policymakers and regulatory authorities collaborate at the international level to harmonize regulatory approaches and standards for AI-enabled IoT systems. This includes participating in forums such as the OECD AI Policy Observatory and the International Telecommunication Union (ITU) to share best practices, exchange information, and develop common frameworks for governing AI and IoT technologies across borders.

Overall, policymakers and regulatory authorities play a central role in governing AI-enabled IoT systems, ensuring that these technologies are developed, deployed, and used in a manner that promotes innovation, protects public interests, and upholds ethical and regulatory standards. By establishing clear policies, regulations, and guidelines, policymakers can create an enabling environment for the responsible and ethical adoption of AI-enabled IoT systems, fostering trust, accountability, and societal well-being.

5.3 *Industry Leaders and Corporations*

Industry leaders and corporations wield considerable influence in shaping the governance landscape of AI-enabled IoT systems as shown in Fig. 6. As pioneers in technological innovation and market drivers, they bear significant responsibility for establishing ethical standards, fostering transparency, and ensuring compliance with regulatory frameworks. One of their primary roles lies in setting industry standards and best practices and collaborating with regulatory bodies and industry associations to develop guidelines that promote interoperability, security, and ethical behavior. Furthermore, these entities play a pivotal role in investing in research and development initiatives, driving technological advancements, and addressing critical challenges such as data privacy and security. By prioritizing ethical design and use, they embed principles of fairness, transparency, and accountability into the development process, ensuring that AI-enabled IoT systems are aligned with societal values and expectations. Additionally, industry leaders engage with stakeholders across various sectors to understand concerns and priorities, fostering trust and collaboration to develop solutions that meet diverse needs. Advocating for responsible policies and demonstrating corporate leadership by integrating ethical practices into governance structures and organizational culture, industry leaders and corporations pave the way for a future where AI-enabled IoT technologies contribute positively to society while upholding ethical principles and safeguarding individual rights.

5.4 *Public Involvement and Awareness*

Public involvement and awareness are critical elements in shaping the governance of AI-enabled IoT systems, ensuring that societal values, concerns, and aspirations are integrated into decision-making processes. As AI and IoT technologies become increasingly pervasive in our daily lives, the public needs to be informed and engaged in discussions surrounding their development, deployment, and use. One of the primary roles of public involvement is to provide stakeholders with opportunities to voice their opinions, share their experiences, and express their concerns regarding AI-enabled IoT systems. This can be achieved through public consultations, town hall meetings, and stakeholder forums, where citizens, advocacy groups, and civil society organizations can actively participate in discussions and provide feedback on issues such as data privacy, algorithmic bias, and ethical considerations. Furthermore, raising public awareness about the capabilities, risks, and societal implications of AI-enabled IoT systems is crucial for fostering informed decision-making and promoting digital literacy. Educational campaigns, public awareness initiatives, and outreach programs can help empower individuals to make informed choices about their use of AI-enabled IoT technologies and advocate for policies that protect their rights and interests. By actively involving the public in governance processes and raising awareness about the opportunities and challenges associated with AI-enabled

Fig. 6 Industry leaders and corporations

IoT systems, policymakers, industry leaders, and other stakeholders can build trust, foster transparency, and ensure that these technologies serve the common good and contribute positively to society.

6 Sector-Specific Governance Frameworks

Sector-specific governance frameworks are pivotal for regulating the deployment and operation of AI-enabled IoT systems within distinct industries. These frameworks are meticulously crafted to address sector-specific challenges, regulations, and objectives, ensuring that AI and IoT technologies are harnessed responsibly to meet industry needs while adhering to ethical and legal standards. For instance, in healthcare, governance frameworks prioritize patient safety, privacy, and data security, aligning with regulations like HIPAA in the U.S. to safeguard sensitive health information. Meanwhile, in finance, frameworks focus on ensuring financial stability and consumer protection, guided by regulations such as GDPR and PCI DSS to uphold data integrity and privacy in financial transactions. Similarly, transportation frameworks emphasize passenger safety and infrastructure resilience, with regulations like FAA guidelines for unmanned aerial systems governing AI-driven aviation technologies. Energy and manufacturing sectors also implement governance frameworks to optimize energy efficiency, grid reliability, and product quality, complying with standards like NERC and ISO to ensure operational excellence and environmental sustainability. By tailoring governance frameworks to sector-specific needs, industries can navigate the complexities of AI-enabled IoT systems effectively, fostering innovation while safeguarding public trust and regulatory compliance. Sector-specific governance frameworks are tailored sets of rules, guidelines, and standards designed to regulate the development, deployment, and use of AI-enabled IoT systems within specific industries or sectors. These frameworks aim to address sector-specific challenges, opportunities, and regulatory requirements, ensuring that AI and IoT technologies are deployed responsibly and effectively to achieve sector-specific objectives. Here are examples of sector-specific governance frameworks in key industries:

- **Healthcare**: In the healthcare sector, governance frameworks focus on ensuring patient safety, privacy, and data security while harnessing the potential of AI-enabled IoT systems to improve medical diagnosis, treatment, and patient care. These frameworks often include regulations such as the Health Insurance Portability and Accountability Act (HIPAA) in the United States, which governs the use and disclosure of patients' health information, as well as industry-specific standards for data interoperability, medical device regulation, and ethical guidelines for AI-driven medical decision-making. Example: A governance framework for healthcare might include specific guidelines for Electronic Health Record (EHR) systems, telemedicine platforms, and medical IoT devices. It would outline protocols for ensuring the security and privacy of patient data, establishing consent mechanisms for data sharing, and integrating with existing healthcare workflows to improve care coordination and patient outcomes.
- **Finance**: Governance frameworks in the finance sector aim to promote financial stability, consumer protection, and data integrity while leveraging AI-enabled IoT systems to enhance risk management, fraud detection, and customer service. Regulations such as the European Union's General Data Protection Regulation (GDPR) and the Payment Card Industry Data Security Standard (PCI DSS) set requirements for data protection and security in financial transactions, while industry guidelines for algorithmic trading and robo-advisors address the ethical and operational challenges associated with AI-driven financial services. Example: A governance framework for financial technology (fintech) platforms might include specific guidelines for digital payment systems, peer-to-peer lending platforms, and cryptocurrency exchanges. It would outline protocols for ensuring the security and integrity of financial transactions, implementing Anti-Money Laundering (AML) and Know-Your-Customer (KYC) procedures, and complying with regulatory requirements related to financial services and consumer protection.
- **Transportation**: In the transportation sector, governance frameworks focus on ensuring passenger safety, infrastructure resilience, and environmental sustainability while integrating AI-enabled IoT systems into transportation networks,

vehicles, and infrastructure. Regulations such as the Federal Aviation Administration (FAA) regulations for Unmanned Aerial Systems (UAS) in the aviation industry and the National Highway Traffic Safety Administration (NHTSA) guidelines for autonomous vehicles in the automotive industry set safety and performance standards for AI-driven transportation technologies. Example: A governance framework for transportation might focus on integrating smart technologies, such as Intelligent Transportation Systems (ITS), connected vehicles, and smart infrastructure, to optimize traffic flow, reduce congestion, and enhance safety. It would outline standards for data exchange between transportation stakeholders, facilitate interoperability among different transportation systems, and address ethical and regulatory considerations related to autonomous vehicles and Mobility-as-a-Service (MaaS) platforms.

- **Energy**: Governance frameworks in the energy sector aim to promote energy efficiency, grid reliability, and environmental sustainability while leveraging AI-enabled IoT systems to optimize energy production, distribution, and consumption. Regulations such as the North American Electric Reliability Corporation (NERC) standards for cybersecurity in the electric grid and the European Union's Energy Efficiency Directive set requirements for data security, interoperability, and energy management systems in the energy industry. Example: A governance framework for smart energy grids might include specific guidelines for smart metering systems, energy management platforms, and Distributed Energy Resources (DERs) such as solar panels and battery storage systems. It would outline protocols for optimizing energy consumption, managing grid congestion, and facilitating peer-to-peer energy trading.
- **Manufacturing**: In the manufacturing sector, governance frameworks focus on ensuring product quality, supply chain resilience, and worker safety while adopting AI-enabled IoT systems for smart manufacturing, predictive maintenance, and quality control. Standards such as the International Organization for Standardization (ISO) 9001 for quality management and ISO 27001 for information security management provide guidelines for implementing AI-driven manufacturing processes while ensuring compliance with regulatory requirements and industry best practices. Example: A governance framework for smart manufacturing facilities might include specific guidelines for industrial automation systems, digital twin technologies, and predictive maintenance solutions. It would outline protocols for optimizing production processes, minimizing downtime, and ensuring product quality. For instance, a smart manufacturing governance framework might mandate the use of digital twin technology to create virtual replicas of physical manufacturing assets, such as machinery, equipment, and production lines. By simulating and analyzing the behavior of these assets in real-time, digital twins enable manufacturers to optimize performance, identify potential issues, and make data-driven decisions to improve efficiency and productivity.

Overall, sector-specific governance frameworks play a crucial role in guiding the responsible and ethical deployment of AI-enabled IoT systems across different industries, addressing sector-specific challenges and opportunities while ensuring compliance with regulatory requirements and industry standards. By adopting these frameworks, organizations can harness the transformative potential of AI and IoT technologies to drive innovation, improve operational efficiency, and deliver value to stakeholders while mitigating risks and safeguarding public interests.

7 Top 5 World Case Studies

- **Smart City Initiatives in Singapore**

Singapore has been at the forefront of implementing AI-enabled IoT technologies to transform into a smart city. To maximize public services, transportation networks, and urban infrastructure, the city-state makes use of IoT sensors, AI algorithms, and data analytics in combination. For instance, the Smart Water Grid uses Internet of Things sensors to monitor water consumption and identify leaks in the water supply network, and the Smart Traffic Lights System uses real-time data to optimize traffic flow and reduce congestion. These projects are part of Singapore's Smart Nation initiative.

- **Precision Agriculture in the United States**

The US has been implementing AI-enabled IoT devices in the agriculture industry to raise crop yields, maximize resource use, and improve farm management techniques. Businesses such as John Deere have created AI-driven agricultural equipment with the Internet of Things sensors to track crop health, soil moisture content, and meteorological factors in real-time. Through data analysis, farmers may make informed decisions about pest management, fertilization, and irrigation, resulting in higher agricultural output and sustainability.

- **Healthcare Innovation in Israel**

Israel has emerged as a leader in healthcare innovation, leveraging AI-enabled IoT systems to enhance patient care, medical diagnosis, and treatment outcomes. Companies like Zebra Medical Vision have developed AI algorithms. These algorithms are useful in medical imaging data analysis, such as X-rays and CT scans, to assist radiologists in detecting diseases like cancer and cardiovascular conditions. These AI-enabled diagnostic tools improve diagnostic accuracy and efficiency, leading to better patient outcomes and reduced healthcare costs.

- **Smart Manufacturing in Germany**

Germany, known for its advanced manufacturing capabilities, has been integrating AI-enabled IoT systems into its industrial processes to drive efficiency, quality, and innovation. Companies like Siemens and Bosch are implementing AI-powered predictive maintenance systems that use IoT sensors to monitor equipment health and predict potential failures before they occur. By proactively addressing maintenance needs, manufacturers can minimize downtime, optimize production schedules, and reduce operational costs.

- **Renewable Energy Integration in Australia**

To more successfully incorporate renewable energy sources like solar and wind power into its energy infrastructure, Australia is utilizing AI-enabled IoT technologies. To stabilize the grid and provide grid balancing services, projects like the Virtual Power Plant (VPP) project in South Australia employ Artificial Intelligence (AI) algorithms to collect and manage distributed energy resources like rooftop solar panels and battery storage devices. Australia hopes to make the shift to a more robust and sustainable energy system by making the most of its renewable energy resources. These case studies illustrate the diverse applications of AI-enabled IoT systems across different sectors and regions, showcasing their potential to drive innovation, improve efficiency, and address societal challenges on a global scale.

7.1 Future Trends and Recommendations

The governance of AI-enabled Internet of Things systems is shaped by several future proposals. First, improving accountability and transparency in the creation and application of these technologies is probably going to remain a top priority. Stakeholders from many businesses will progressively want openness regarding AI algorithms, data use, and decision-making procedures to guarantee equity, reduce prejudices, and foster user confidence. Furthermore, there will be an increasing emphasis on putting in place strong governance and regulatory frameworks that are specifically designed to handle the benefits and particular issues that AI-enabled IoT systems provide. These frameworks should prioritize data privacy, security, and ethical considerations while fostering innovation and competitiveness.

Furthermore, as AI and IoT technologies become more integrated into critical infrastructure and essential services, there will be a heightened emphasis on ensuring resilience and reliability in the face of emerging threats and disruptions. This will require proactive risk management strategies, including the development of cybersecurity protocols, incident response plans, and continuity measures to mitigate the impact of cyberattacks, system failures, and natural disasters. In terms of recommendations, policymakers, industry leaders, and other stakeholders should collaborate closely to develop and implement comprehensive governance frameworks that promote responsible and ethical AI-enabled IoT deployments. This includes fostering cross-sectoral partnerships, sharing best practices, and promoting international cooperation to address common challenges and regulatory gaps. Additionally, there should be a concerted effort to invest in research and development initiatives aimed at advancing AI and IoT technologies while ensuring they align with societal values, ethical norms, and human rights principles. Moreover, stakeholders should prioritize efforts to enhance digital literacy and promote public awareness about the capabilities, risks, and implications of AI-enabled IoT systems. This includes educating users about their rights, responsibilities, and options for controlling their data, as well as fostering a culture of data stewardship and responsible innovation. Empowering people to make informed choices and engage in governance processes helps guarantee that AI-enabled IoT systems serve the common good and improve society while protecting ethical values and individual rights and freedoms.

Moving forward, there are several key recommendations for future research and implementation efforts in the realm of AI-enabled IoT systems. Firstly, there is a critical need for interdisciplinary research that bridges the gap between technical expertise and domain-specific knowledge. Collaborative efforts involving experts from fields such as computer science, engineering, social sciences, ethics, and law can help address complex challenges related to privacy, security, fairness, and accountability in AI-enabled IoT deployments. Secondly, there is a pressing need for research that explores the long-term societal impacts of AI-enabled IoT systems. This includes studying their implications for employment, economic inequality, healthcare access, environmental sustainability, and democratic governance. Understanding these broader societal implications is essential for developing governance frameworks that promote equitable and sustainable outcomes for all stakeholders. Thirdly, there is a need for research that focuses on addressing the ethical and human rights implications of AI-enabled IoT systems. This includes developing frameworks for assessing the ethical implications of AI algorithms, ensuring transparency and accountability in algorithmic decision-making, and protecting fundamental rights such as privacy, freedom of expression, and non-discrimination. Additionally, there is a need for research that explores innovative approaches to cybersecurity and data protection in AI-enabled IoT systems. This includes developing new encryption techniques, authentication mechanisms, and intrusion detection systems to safeguard against cyber threats and data breaches. Moreover, there is a need for research that investigates the potential of decentralized and federated AI models to enhance data privacy and security in IoT environments.

Additionally, studies assessing the efficiency of regulatory frameworks and governance structures in guaranteeing the ethical and responsible creation and use of AI-enabled Internet of Things systems are required. Policymakers can better handle new possibilities and difficulties in the quickly changing world of AI-enabled IoT devices by assessing the advantages and disadvantages of current governance frameworks and making necessary improvements.

8 Conclusion

In conclusion, the governance of AI-enabled IoT systems represents a complex and multifaceted challenge that requires collaborative efforts from policymakers, industry leaders, researchers, and other stakeholders. It is essential to develop robust governance frameworks that promote responsible and ethical deployment while safeguarding individual rights, public interests, and societal values. Transparency, accountability, data privacy, cybersecurity, and stakeholder engagement are some of the important components of governance in AI-enabled IoT systems that have been discussed throughout this debate. Researchers have highlighted case studies demonstrating the varied uses of AI-enabled IoT technology across businesses and locations, as well as have investigated sector-specific governance structures. It is critical to focus future research and implementation efforts on AI-enabled IoT system governance by anticipating and responding to new possibilities and threats. Research on the social effects of these technologies in the long run, new methods of cybersecurity and data protection, and multidisciplinary studies that bring together technological and ethical concerns are all part of this. Furthermore, to find ways to make things better, it is essential to assess how well current regulatory frameworks and governance structures are working. Everyone can drive innovation, improve efficiency, and address societal challenges with AI-enabled IoT systems by collaborating on comprehensive governance frameworks. These frameworks will respect individual rights and freedoms, uphold ethical principles, and enable us to harness the transformative potential of these systems. People can make sure that AI-enabled IoT technologies help people and the planet in the future by continuing to work together as a governing community. In conclusion, the significance of customizable governance frameworks that harmonize technical innovation with ethical governance cannot be overstated. To navigate the complexities of integrating smart systems into various aspects of society, these frameworks serve as guiding beacons, ensuring responsible deployment and utilization. Moving forward, it is imperative that it not only acknowledges the importance of these frameworks but also actively engages in their continued development and implementation. This requires a collective commitment from stakeholders across academia, industry, government, and civil society to collaborate, innovate, and address emerging challenges head-on.

References

1. Mishra, M.K.: Role of Technology in Smart Governance: 'smart City, Safe City, Safe City' (2013)
2. Pereira, G.V., Parycek, P., Falco, E., Kleinhans, R.: Smart governance in the context of smart cities: a literature review. Inf. Polity **23**, 143–162 (2018)
3. Vinod Kumar, T.M.: E-Governance for Smart Cities. Springer, Singapore (2015)
4. Andersen, D.L., Ashbrook, C.S.A., Karlborg, N.B.: Significance of big data analytics and the internet of things (IoT) aspects in industrial development, governance and sustainability. Int. J. Intell. Netw. **1**, 107–111 (2020)
5. Ismagilova, E., Hughes, L., Rana, N.P., Dwivedi, Y.K.: Security, privacy and risks within smart cities: literature review and development of a smart city interaction framework. Inf. Syst. Front. 1–22 (2022)
6. Pagallo, U., Aurucci, P., Casanovas, P., Chatila, R., Chazerand, P., Dignum, V., Luetge, C., Madelin, R., Schafer, B., Valcke, P.: On good AI governance: 14 priority actions, a SMART model of governance, and a regulatory toolbox (2019)
7. Chianese, A., Piccialli, F.: A smart system to manage the context evolution in the cultural heritage domain. Comput. Electr. Eng.. Electr. Eng. **55**, 27–38 (2016)

8. Rottondi, C., Verticale, G., Fraternali, P., Novak, J., Pelechrinis, K., Hidasi, B., Vaca, C.: IEEE Access Special Section Editorial: Social computing applications for smart cities. IEEE Access **7**, 65219–65222 (2019)
9. Milchram, C., Künneke, R., Doorn, N., van de Kaa, G., Hillerbrand, R.: Designing for justice in electricity systems: a comparison of smart grid experiments in the Netherlands. Energy Policy **147** (2020)
10. Darmawan, A.K., Siahaan D, Susanto, T.D., Hoiriyah, H., Umam, B., Anwari, A.: Exploring success factor for mobile based smart regency service using TRUTAUT model approach. In: 7th IEEE International Conference on Electrical Engineering, Computer Sciences and Informatics, pp. 250–256 (2021)
11. Fiorentino, S., Bartolucci, S.: Blockchain-based Smart contracts as new governance tools for the sharing economy. Cities **117** (2021)
12. Biyik, C., Allam, Z., Pieri, G., Moroni, D., O'Fraifer, M., O'Connell, E., Olariu, S., Khalid, M.: Smart parking systems: reviewing the literature, architecture and ways forward. Smart Cities **4**, 623–642 (2021)
13. Taratori, R., Rodriguez-Fiscal, P., Pacho, M.A., Koutra, S., Pareja-Eastaway, M., Thomas, D.: Unveiling the evolution of innovation ecosystems: an analysis of triple, quadruple, and quintuple helix model innovation systems in European case studies. Sustainability (2021)
14. Hentea, M.: Distributed energy systems (2021)
15. Lakhwani, K., Sharma, G., Sandhu, R., Nagwani, N.K., Bhargava, S., Arya, V., Almomani, A.: Adaptive and convex optimization-inspired workflow scheduling for cloud environment. Int. J. Cloud Appl. Comput. (IJCAC) **13**(1), 1–25
16. Hine, C., Nilforooshan, R., Barnaghi, P.: Ethical considerations in design and implementation of home-based smart care for dementia. Nurs. Ethics. Ethics **29**, 1035–1046 (2022)
17. Sandhu, R., Faiz, M., Kaur, H., Srivastava, A., Narayan, V.: Enhancement in performance of cloud computing task scheduling using optimization strategies. Cluster Comput. 1–24 (2024)
18. Lane, L.: Preventing long-term risks to human rights in smart cities: a critical review of responsibilities for private AI developers. Internet Policy Rev. **12** (2023)
19. Peter, M.: Embedded System Design: Embedded Systems Foundations of Cyber-Physical Systems, and the Internet of Things. Springer, Berlin (2021)
20. Ramandeep, S., Lakhwani, K.: Improved scientific workflow scheduling algorithm with distributed heft ranking and TBW scheduling method. In: International Conference on Wireless Sensor Networks, Ubiquitous Computing and Applications, vol. 244 (2022). https://doi.org/10.1007/978-981-16-2919-8
21. Ahram, T.Z., Karwowski, W., Amaba, B.: User-centered systems engineering & knowledge management framework for design & modeling of future smart cities, In: Proceedings of the Human Factors and Ergonomics Society Annual Meeting, vol. 54, no. 20, pp. 1752–1756. SAGE Publications, Los Angeles (2010)
22. Channi, H.K., Handa, K., Bhullar, M., Sandhu, R., Kaur, H., Singh, P.: Renewable distributed generation for smart highway lighting. In: 2024 11th International Conference on Reliability, Infocom Technologies and Optimization (Trends and Future Directions) (ICRITO). IEEE, pp. 1–6 (2024)
23. Deepika, G., Tripathi, S.L., Saxena, S., Chanda, M., Alazab, M. (eds.): Machine Learning Algorithms for Signal and Image Processing. Wiley, London (2022)
24. Ghai D, Gianey HK, Jain A, Uppal RS (2020) Quantum and dual-tree complex wavelet transform-based image watermarking. Int. J. Mod Phys. B **34**, 2050009 (2020)
25. Sandeep, K., Deepika, G., Arpit, J., Tripathi, S.L., Shilpa, R.: Multimodal Biometric and Machine Learning Technologies: Applications for Computer Vision. Wiley, London (2023)
26. Sandhu, R., Channi, H.K., Ghai, D., Cheema, G.S., Kaur, M.: An introduction to generative AI tools for education 2030. In: Integrating Generative AI in Education to Achieve Sustainable Development Goals, pp. 1–28 (2024)
27. Raju, B.A.N., Deepika, G., Tripathi, S.L., Nanda S.K., Islam, S.M.N.: Predictive analytics for marketing and sales of products using smart trolley with automated billing system in shopping malls using LBPH and faster R-CNN. Intell. Techn. Predictive Data Anal. **111** (2024)

Embracing the Synergy of Cloud Computing and Business Intelligence

Riya Bhatia, Ananyaa Sharma, and Dolly Sharma

Abstract In the dynamic landscape of modern business, the convergence of cloud computing and business intelligence (BI) emerges as a transformative force. This chapter explores the seamless integration of cloud computing and BI, unveiling their collaborative potential to reshape organizational strategies and decision-making processes. From scalable data storage in the cloud to real-time analytics capabilities, this exploration encompasses the multifaceted aspects of harnessing cloud technologies for robust BI solutions. By delving into real-world examples and practical applications, the chapter navigates the symbiotic relationship between cloud computing and BI, offering insights into how businesses can embrace this synergy to drive innovation, enhance operational efficiency, and gain a competitive edge in today's data-driven era.

1 Introduction

Before cloud computing came into existence, several types of computing were there. First of all Client–Server Computing was used which is a type of centralized storage where all the data and software controls are restricted to the server side only. Now is a user needs access over a data, then he/she first needs to connect to the server for successful access. But the server gets overloaded if the clients request the access over data and other controls at the same time. So, after the Client–Server computing, Distributed computing came into view. In Distributed Computing, all the computers are connected together which share resources among themselves to solve a problem or perform a task. But it also had certain limitations such as it needed highly skilled labour due to its complex design and it wasn't cost efficient due to high maintenance. So then 'Cloud Computing' came in picture which shaped the world by transforming

R. Bhatia · A. Sharma · D. Sharma (✉)
Department of Computer Science and Engineering, Amity University, Noida, India
e-mail: dolly.azure@gmail.com

S. Mahajan et al. (eds.), *Smart Systems: Engineering and Managing Information for Future Success*, Information Systems Engineering and Management 22,
https://doi.org/10.1007/978-3-031-76152-2_12

or revolutionizing the industry by providing simplified resource management, cost-effective scalability, automated maintenance and never known and unmatched flexibility in resource allocation and utilization. These advantages of Cloud Computing attracted researches to explore its potential in Business Intelligence (or BI). This led to the integration of Cloud Computing and Business Intelligence and 'Cloud BI' emerged as a solution in harnessing the advantages of the cloud for data analysis and decision-making processes [1].

2 Understanding Cloud Computing

2.1 What Is Cloud Computing

Cloud Computing is the on-demand technology/delivery of IT resources via the internet that allows us to create, update or modify various applications. In simpler terms, Cloud Computing enables access to computing services such as servers, storage, databases, networking, software, and more over the internet, without the need for physical hardware or infrastructure. This approach replaces the traditional model of owning and maintaining physical data centers with a more flexible and scalable solution. Cloud Computing data centres are managed by Cloud Service Provider (or CSP) for example Google Cloud, Microsoft Azure, Amazon Web Services (AWS), etc. [2].

Cloud services are typically offered on a pay-as-you-go basis, providing organizations with cost-effective solutions tailored to their needs. This means that instead of investing heavily in upfront infrastructure costs, organizations can scale their IT resources up or down based on demand, paying only for what they use. Additionally, Cloud Computing offers benefits such as enhanced reliability, security, and accessibility. Data stored in the cloud can be accessed from anywhere with an internet connection, promoting collaboration and remote work.

IT companies do face some challenges at smaller or larger levels like they need Server room which is the basic need for any company. Then these server rooms should be equipped with database servers, mail servers, networking equipment, firewalls, routers, modems, switches, and high-speed internet connectivity. These systems also need high maintenance which leads to the requirement of highly-skilled labour which further leads to economic burden. These challenges eradicated with the advent of 'Cloud Computing'. Instead of maintaining physical data centres, companies can harness cloud services and fulfil their requirements. The eradication of physical data centres will lead to reduction in costs and increase in flexibility [3].

In essence, Cloud Computing can be likened to renting a house instead of building one from scratch. Instead of investing in land, materials, and construction, one can simply rent a fully furnished property, tailored to their needs, for as long as required. Similarly, CLOUD COMPUTING offers access to IT resources, eliminating the need

Fig. 1 Cloud computing

for upfront investment in hardware and infrastructure, while providing scalability and flexibility (Fig. 1).

2.2 Types of Cloud Computing

As the digital landscape evolves, the diversity of cloud computing models plays a pivotal role in shaping how businesses leverage technology. The classification of cloud computing into distinct types reflects the versatility and tailored solutions available to organizations. This section explores the various types of cloud computing, each with its unique characteristics, to provide insights into the strategic choices that define modern IT infrastructures.

- Public Cloud: Services provided by third-party cloud service providers and these services can be accessed by the users (public) over the internet.
- Private Cloud: Services are provided and managed by the third party Cloud Service Providers or the organization itself but are dedicated to and accessible by single customer only.
- Hybrid Cloud: A combination of public and private cloud framework, where organizations can harness the benefits of both cloud environments (Fig. 2) [4].

2.3 Cloud Service Models

In the dynamic realm of cloud computing, understanding the diverse service models is fundamental to crafting an effective IT strategy. Cloud service models delineate the

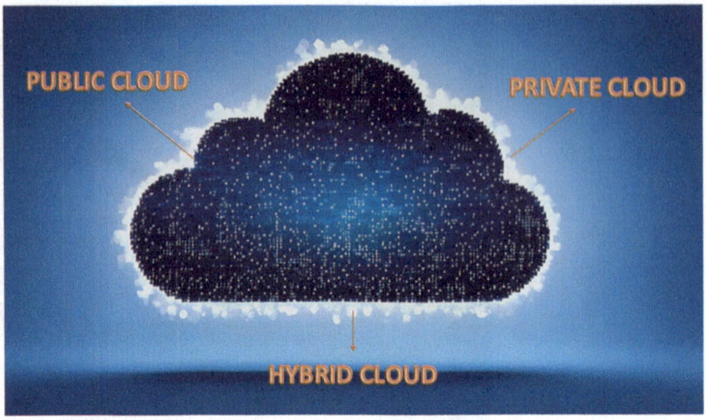

Fig. 2 Types of cloud computing

roles and responsibilities between service providers and users, offering a spectrum of solutions to meet specific organizational needs. This section delves into the nuances of cloud service models, providing a foundational understanding of Infrastructure as a Service (IaaS), Platform as a Service (PaaS), and Software as a Service (SaaS) to navigate the evolving landscape of digital infrastructure.

- Infrastructure as a Service (IaaS): In IaaS, users rent virtualized hardware resources such as like servers, storage, machine and networking components from the Cloud Service Providers. Without depending on physical hardware's, users can control their framework easily.
- Platform as a Service (PaaS): PaaS offers a platform where users can build, run, manage and monitor their applications. User don't need to worry about the underlying framework. It offers pre-configured platform which has advanced tools and services for the different development processes.
- Software as a Service (SaaS): In SaaS, User have to buy a subscription to use software applications over the internet. Users don't need to install the underlying framework to use the applications. They can directly access them over the internet through web browsers and client applications (user interfaces). And the Cloud Service Providers manage, maintains and updates the software itself.

3 Understanding Business Intelligence (BI)

3.1 What Is Business Intelligence (or BI)

Business Intelligence is a software that collects and analyses business data to present it in user-friendly format such as graphs, reports, charts, etc. BI serves as a powerful tool for decision-making and strategic planning within organizations. It enables users to gain insights into their operations, identify trends, and make informed decisions to drive growth and efficiency. In simple words it is combination of software tools and hardware framework/infrastructure that has the aim to provide help in strategic planning of an organization. Different platforms of BI include Google Data Studio, MicroStrategy, SAP Business Objects, IBM Cognos, Tableau, etc. These platforms provide with BI tools which help in creating reports, dashboards, charts, graphs, etc. [5].

In practical terms, BI can be compared to a dashboard in a car. Just as a dashboard provides essential information to the driver, such as speed, fuel levels, and engine status, BI tools offer key metrics and performance indicators to business users, allowing them to monitor the health of their organization and make course corrections as needed. BI solutions aggregate data from various sources within an organization, including databases, spreadsheets, and software applications, providing a comprehensive view of business performance.

BI platforms offer a range of functionalities, including data visualization, ad-hoc reporting, predictive analytics, and data mining. These tools empower users to explore data, uncover insights, and communicate findings effectively across departments. By leveraging BI, organizations can streamline operations, optimize processes, and gain a competitive edge in the marketplace.

3.2 *Data Framework for BI*

Data framework of BI refers to the structure and organization of data that is used for the processing and analytics of BI. The processes which make up the framework for Business Intelligence are Data Sources, Data Integration, Data Warehousing, Data Modeling, Data Governance, Data Analysis and Reporting [6].

3.3 *Types of BI*

In the multifaceted field of Business Intelligence (BI), recognizing the diverse types of BI applications is crucial for organizations seeking to harness actionable insights. These variations in BI types cater to distinct business needs and analytical requirements, offering a spectrum of tools and approaches. This section explores the landscape of BI, encompassing operational, strategic, and tactical BI, to provide a comprehensive understanding of the multifunctional role BI plays in organizational decision-making.

- Self-Service BI: Self-Service BI empowers users can independently access the tools and features without the need for technical guidance.
- Mobile BI: In Mobile BI, Real-time data, different tools and features can be accessed on mobile device. This helps users to take decisions on-the spot if needed.

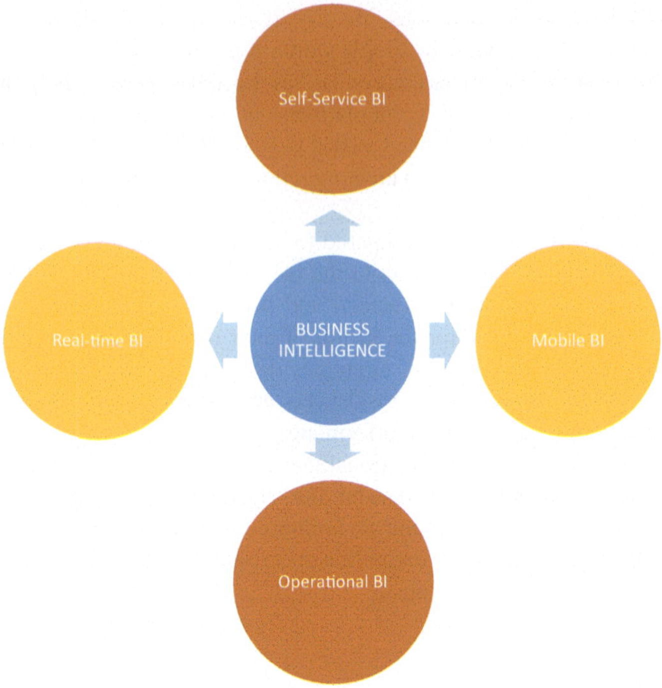

Fig. 3 Types of business intelligence

- Real-time BI: It offers data which is updated and analysed per minute for nimble and proactive decision-making which in turn benefits the organization or the user.
- Operational BI: Operational BI leverages BI for monitoring the Real-Time data and further optimizing it as per requirement (Fig. 3) [7].

4 Importance of Integrating Cloud Computing and Business Intelligence

Integrating cloud computing and BI will lead to the almost complete eradication of obsolescence (out datedness) problem of technology. Merging Cloud BI will help an organization by increasing its flexibility and scalability, increasing cost efficiency, enabling Real-Time Data Analysis, increasing accessibility, efficient collaboration and decision-making. Also, there is enhanced data security, advancement in the data analysis process and fast deployment and updating speed [8].

5 Understanding Cloud BI Solution

5.1 What Is Cloud BI

When we merge cloud computing and BI, we collectively call them as 'Cloud BI' or 'Cloud BI solutions' or 'Cloud-based BI'. Cloud BI refers to the use of cloud computing in Business Intelligence (BI) which further enhances businesses to produce valuable insights. BI requires large capital layout and processing power to help in the development of Businesses. These requirements often put pressure on these businesses. The Cloud BI solution will help companies to lower the cost of having a BI. Cloud BI will also help companies to have access over latest software's which will benefit the companies by providing an edge on their Competitors (Fig. 4) [9].

In short, Cloud BI has the ability to become a lynch-pin for various enterprises by providing end-users with real-time access to business-critical data [10].

Fig. 4 Cloud BI solutions

5.2 Steps Involved in the Process of Integration of Cloud BI

Efficient integration of Cloud Business Intelligence (BI) into organizational workflows demands a strategic and methodical approach. The seamless amalgamation of cloud-based BI solutions involves a series of well-defined steps aimed at optimizing data accessibility and analytical capabilities. In this section, we explore the key steps in the integration process, offering insights into how organizations can navigate the complexities of adopting and integrating Cloud BI for enhanced decision-making.

- Requirement Gathering: Make a clear list or jot down your clear goals to coordinate with your business metrics or Key Performance Indicators (KPI's).
- Data Source Identification: Identify various data sources like databases, spreadsheets, CRM systems, etc. that need to be integrated with Cloud BI to ensure data quality and maintenance.
- Data Extraction and Transformation: This process helps in extracting data from various sources and transforming it into harmonized format.
- Data Loading and Storage: Load the data on the cloud data warehouses where data can be analysed.
- Data Modeling and Schema Design: Design the data model and schema for efficient data storage and recovery.
- Data Security and Governance: Implement necessary security measures to protect data and ensure that the data is in compliance with the government standards and privacy regulations.
- BI Tool Configuration: Configure or customize the Cloud BI platform or tools according to organizational requirements.
- Testing and Validation: Meticulously test the integrated Cloud BI system (check for the proper working of tools and various functions) and verify the results against benchmarks.
- User Training and Adoption: Cloud BI has a complex framework, so provide training facilities to the user to acquaint the user with various tools and features for enhanced user adoption which in turn will increase productivity of the organization.
- Ongoing Maintenance and Monitoring: Consistently supervise the Cloud BI system by regularly performing updates, and ensuring data scalability [11].

5.3 Advantages of Cloud BI

- Low maintenance cost or cost effectiveness: Cloud BI minimizes both hardware and software maintenance costs for organizations.
- Data Security: Cloud BI follows appropriate protocols to protect the data of user which is being delivered over the cloud. Data is safeguarded from being breached by implementing advanced techniques like encryption algorithms and very strong access controls.

- Improved Collaboration: Cloud Computing provides an option of shared storage where users can easily group themselves and collaborate to share data.
- Scalability and Accessibility: Cloud BI has high scalability, as it allows businesses to increase or decrease the resources as per requirement without effecting the framework. Cloud BI also has high accessibility and flexibility as users can access the data from any device, anywhere (provided a stable internet connection) [12].
- Enhanced decision-making: Cloud BI offers valuable insights which further helps businesses to take better and informed decisions, which leads to improved outcomes
- Enhanced data visualization: In Cloud BI, BI tools offer visually appealing front-end that helps in understanding complex data more easily and efficiently.
- Cutting edge mobility: In Cloud BI, Cloud computing allows us to access data very easily 24/7/365.25 anywhere using any device which is internet active. Users are not bound to a physical location which results in increased productivity and effective collaboration.
- Online and unlimited Data Storage: Cloud Computing allows us to store, analyse and access data like audio, video, files, etc. on cloud storage. There is no need to maintain physical data centres for storing huge amounts of data as they require a lot of space and are too costly. So as Cloud Computing offers these benefits, so indirectly Cloud BI offers this benefits as well. Cloud BI provides and easy to use interface with unlimited storage capacity and has also provides high speed Example: Box.com is a data storage application which allows user to drag and drop the files into box and access the file from anywhere.
- Backup and Recovery: In the conventional system backing up the data is a complex problem and in case of a disaster, there may be complete loss of data. Cloud Service Providers or CSP provides safe backup and recovery options and in case of a disaster, data can be recovered with miniscule damage. Example: Mozy is backup and recovery application which allows user to backup their data and recover it if the data gets lost.
- Pay-per-user model: Pay-per-user model also known as pay-as-you-go approach is a pricing structure where users are charged based on the number of individuals using a particular service. This model leads to cost optimization and increases productivity and flexibility. Cloud BI also provides Application Programming Interfaces (APIs) to the users which helps in accessing cloud services and ensure the accurate billing in the pay-per-user model [1].

5.4 Disadvantages of Cloud BI

- Dependence on internet connectivity: Cloud BI needs stable internet connection so if the internet speed hinders, then it may lead to interruption in access to cloud services.
- Potential outage: There may be ephemeral or temporary unavailability of services due to potential downtime in services.

- Limited customization options: There is limited control over the framework of cloud for the users which resists or hinders Customization options.
- Vendor Lock-in or Migration challenges: Migrating to different Cloud Service Providers is complex and time consuming and is very expensive due to vendor-specific technologies and data mobility.
- Heavy dependence on Cloud Service Providers(CSPs): When organizations become too much dependent on Cloud Service Providers then it becomes difficult for them to migrate to different CSPs so this leads to decrease in flexibility and scalability and increases complexity in the transition to other CSPs.
- Bandwidth Limitations: If the internet connection used by organizations has bandwidth limitations (which is a very common problem) then the data transfer speed will be low and transferring large amounts of data will become difficult and time-consuming resulting in decreased productivity and flexibility.
- Compliance and regulatory issues: Compliance issues refer to the challenges that organizations face if their data is not in compliance with the government standards and privacy laws and regulations. If organizations are not able to counter these compliance issues, then it may lead to the damage of their reputation and can cause them to land in various legal cases.
- Complexity and Cost: Skilled labour, resources, advanced tools and techniques are required as Cloud BI solutions can be complex and costly.
- Time-consuming implementation: Cloud BI setup can be time-consuming, as it its setup includes many complex processes like data integration, transformation, and data modeling.
- User adoption and training: Cloud BI solutions need to provide training facilities to the users as well as the management team as it has a complex framework. So for improved data-driven decision-making and increased productivity effective user and management adoption is required [13].

5.5 Applications of Cloud BI

In the era of data-driven decision-making, the applications of Cloud Business Intelligence (BI) serve as catalysts for organizational growth and efficiency. Cloud BI transcends traditional boundaries, empowering businesses to extract valuable insights from data seamlessly. This section explores the varied applications of Cloud BI, showcasing how it enhances reporting, analytics, and decision-making processes in today's dynamic business landscape.

- Big Data Analysis: Big Data means extremely large sets of data that cannot be easily examined and handled using conventional methods. Immense volume of Big Data makes it almost impossible to store it using conventional methods. Cloud BI helps in storing this large amount of data as it has unlimited storage capacity and also this Big Data is analysed in a centralized data warehouse. Also there is no need to maintain costly and capacious physical data centres as one can store data on the cloud online. This helps organizations to store and analyse larger sets of Big Data to produce valuable insights.

- Real-Time Analytics: Cloud BI offers real-time data analysis and reporting, which is enables organizations to make data-driven decision-making. Real-time analytics aids in tracking metrics, spotting trends, and uncovering optimization prospects [14].
- Customer Relationship Management (CRM): Cloud BI helps businesses to harness valuable insights like user preferences and behavioural patterns by integrating, analysing and monitoring customer's/users data or data pooling from various sources. By gaining these insights businesses can improve personalized experiences of users and can change marketing strategies as per requirements, thus improving the productivity of the organization [4].
- E-Commerce Applications: E-Commerce refers to the buying and selling of products online. Cloud BI helps e-businesses to respond to opportunities quickly and efficiently by providing large storage size and fast, flexible and easy to handle interface [15].
- Data Integration and ETL: Cloud BI helps in enhancing the process of data integration and ETL processes (Extraction, Transformation, Loading). ETL processes further help in extracting data from various sources, transforming it into harmonized format, and loading it on the cloud data warehouses (for the purpose of data analysis) [16].

5.6 Real-World Examples of Businesses Using Cloud BI as a Solution

In the ever-evolving landscape of business technology, real-world examples stand as testament to the transformative impact of Cloud Business Intelligence (BI). Businesses across industries are harnessing the power of Cloud BI to gain competitive advantages, streamline operations, and extract actionable insights from their data. This section highlights compelling instances of successful Cloud BI implementation, illustrating how forward-thinking enterprises leverage this solution to drive innovation and achieve strategic goals.

- Netflix: Netflix uses cloud BI to examine user behaviour, pattern and preferences which will in turn improve overall user experience. By understanding how users interact with their platform, Netflix can tailor its content recommendations, improve content delivery, and enhance the overall user experience. For example, Netflix uses data insights to recommend personalized content, resulting in increased user engagement and satisfaction.
- Airbnb: Airbnb uses cloud-based BI to obtain valuable insights about the user, market and pricing optimization. It will further help Airbnb to know user preferences which will help them in knowing favourite user destinations and accordingly they can optimize their pricing. By analyzing data from user interactions and

market dynamics, Airbnb can optimize pricing strategies, identify popular destinations, and enhance the overall booking experience for both hosts and guests. This data-driven approach allows Airbnb to stay competitive in the dynamic hospitality industry.

- Uber: Uber, a worldwide rides providing platform, enhances overall rider/user experience by using cloud BI. By analyzing large volumes of real-time data, Uber can optimize driver allocation, predict rider demand, and improve service reliability. For example, Uber uses data insights to optimize driver routes, reduce wait times, and enhance the efficiency of its transportation network, ultimately leading to greater customer satisfaction and loyalty.
- Coca-Cola: Coca-Cola gains valuable insights about consumer preferences by harnesses cloud BI. By analyzing data from various sources, including sales figures and consumer feedback, Coca-Cola can identify emerging trends, launch targeted marketing campaigns, and optimize product offerings to meet consumer demand. This data-driven approach enables Coca-Cola to stay ahead of competitors and maintain its position as a market leader in the beverage industry.
- Spotify: Spotify, a renowned music streaming platform, harnesses cloud BI to examine user preferences. By analyzing data on listening habits, song preferences, and user interactions, Spotify can generate personalized music recommendations, curate playlists, and enhance the overall listening experience for its users. This data-driven approach helps Spotify attract and retain users, increase engagement, and drive revenue growth in a competitive market.
- Amazon: Amazon, a leading e-commerce company, harnesses cloud BI to examine user preferences and behaviour along with company's sales figures [17]. By leveraging data insights, Amazon can optimize product recommendations, personalize the shopping experience, and forecast demand for various products. This data-driven approach enables Amazon to make informed decisions, improve customer satisfaction, and drive sales growth across its e-commerce platform.

6 Comparison of Cost Implications: Traditional Approach Versus Cloud-Based BI Approach

Understanding the cost implications of adopting different Business Intelligence (BI) approaches is crucial for organizations seeking to maximize their analytical capabilities while optimizing expenditure. The cost implications between traditional and cloud BI solutions can differ considerably.

Traditional BI solutions:

- In traditional BI solutions, hardware, significant preliminary investments such as software licenses, framework setup, and upkeep are made by the companies.
- These costs can be considerable and may involve expenditure for servers, storage, networking devices, and skilled IT staff to manage and maintain the framework.

- Additionally, enterprises may need to intermittently update, upgrade and substitute hardware and software components, which in turn will lead to more and more expenses [18].

Cloud-based BI solutions:

- On the other hand, cloud-based BI solutions function on a subscription or pay-as-you-go model, which can ensure financial benefits.
- Enterprises can completely ignore or eradicate the preliminary investment outlay on framework and hardware devices.
- They can expand or reduce their resources as per requirement, remunerating only for the services and storage deployed. Cloud-based BI solutions regularly offer managed services, eradicating the burden or load on in-house IT personnel.

Nevertheless, it is crucial to contemplate sustained operational expenditure when comparing traditional and cloud BI solutions. While cloud BI may have lower upfront outlay, the periodic subscription fees can accumulate over time. In addition to that, data transfer and storage outlays must be considered, as they can differ hinging on the volume of data processed and stored in the cloud.

Ultimately, the cost consequences hinge on various factors such as the scope of the cloud BI implementation, particular requirements, and the enterprise's IT strategy. Conducting a meticulous cost analysis, taking into account both preliminary and postliminary expenditures, is crucial to determine the most cost-effective solution or budget-friendly alternative for a particular enterprise [19].

7 Economic Impact of Cloud Computing on BI

The convergence of cloud computing and business intelligence has reshaped the economic landscape for organizations worldwide. The synergy between these two domains not only transforms how data is processed and analyzed but also bears a profound influence on the economic dynamics of businesses. This section delves into the economic ramifications of leveraging cloud computing in the realm of business intelligence, exploring cost implications, efficiency gains, and the overarching financial impact.

- Business Agility: Cloud BI provides nimble data-driven decision making and real-time enabling enterprises to make informed decisions rapidly.
- Global Access to IT infrastructure: Cloud BI platforms ensure access to data and analytics, anywhere and anytime enabling enterprises to harness resources and expertise globally.
- Return on Investment (ROI): ROI helps in the evaluation of Cloud BI by considering factors such as cost efficiency, enhanced data-driven decision making, increased productivity and competitive benefit. Cloud-based business intelligence solutions can be evaluated using ROI analysis helps in assessing the financial advantages and value created by the implementation.

- Increased operational efficiency: Cloud BI optimizes processes, mechanize reporting, and offers real-time access to crucial information, ensuring companies to streamline their operations and resource management.
- Improved performance monitoring: Cloud BI offers enterprises to track and monitor key performance indicators (KPIs), recognize trends, and assess progress towards goals, enables initiative performance management [20].

8 Security and Privacy Considerations in Cloud-Based BI

As organizations embrace the power of cloud-based Business Intelligence (BI), the paramount concerns of security and privacy come to the forefront. Organizations may encounter potential challenges during implementation. Navigating the digital landscape demands a vigilant approach to safeguarding sensitive data and ensuring privacy compliance. This section sheds light on the critical considerations surrounding security and privacy in the context of cloud-based BI, exploring the measures and strategies essential for a secure and trusted BI environment.

- Security Challenges in Cloud BI: Embracing cloud-based BI introduces a spectrum of security challenges that demand meticulous attention and proactive measures. Primarily, organizations must grapple with the heightened risk of security breaches, unauthorized access, and insider threats inherent in the cloud environment.
- Security Breaches: Cloud BI systems are susceptible to various forms of cyberattacks, including but not limited to DDoS attacks, phishing attempts, and malware injections. These breaches can compromise sensitive business data, leading to financial losses, reputational damage, and legal ramifications.
- Unauthorized Entry: The distributed nature of cloud infrastructure introduces vulnerabilities that can be exploited by unauthorized users attempting to gain access to confidential information. Weak authentication mechanisms, inadequate access controls, and misconfigured permissions pose significant threats to the integrity and confidentiality of data stored in cloud BI platforms.
- Insider Risks: Insider threats, whether intentional or unintentional, represent a significant concern for organizations leveraging cloud BI. Employees, contractors, or partners with privileged access may inadvertently expose sensitive data through negligent actions or intentionally exploit vulnerabilities for personal gain or malicious purposes.
- Regulatory Compliance: Compliance with government regulations and industry standards adds another layer of complexity to cloud BI security. Organizations must navigate a labyrinth of data protection laws, such as GDPR, HIPAA, and CCPA, to ensure that their data management practices align with legal requirements and regulatory frameworks.
- Encryption and Data Protection: Execute strong encryption techniques to safegaurd data both at rest and in transit. Use powerful encryption algorithms and secure key administration practices to protect sensitive information.

- Data Segregation and Isolation: Proper data segregation and isolation within the cloud environment should be ensured by harnessing virtual private clouds (VPCs) and firewall rules separate or segregate data of one user from the other user.
- Data Privacy Concerns and Compliance Regulations: Protecting data privacy and ensuring compliance with regulatory mandates are paramount considerations in the implementation and operation of cloud-based BI solutions. Failure to address data privacy concerns and comply with applicable regulations can result in severe consequences, including hefty fines, legal penalties, and irreparable damage to organizational reputation.
- Sensitive Data Exposure: Cloud BI environments store vast amounts of sensitive business data, ranging from customer information and financial records to proprietary insights and intellectual property. Safeguarding this data against unauthorized access, disclosure, or misuse is imperative to maintain data privacy and protect stakeholders' interests.
- Data Sovereignty: Data sovereignty laws dictate where organizations can store and process data, posing challenges for cloud BI deployments that span multiple geographic regions. Ensuring compliance with data residency requirements and navigating jurisdictional complexities are essential considerations for organizations operating in regulated industries or serving customers across international borders.
- Privacy by Design: Adopting a privacy-by-design approach is essential for embedding data privacy principles into the design, development, and deployment of cloud BI solutions. Implementing privacy-enhancing technologies, anonymization techniques, and robust access controls can help mitigate privacy risks and foster trust among users and stakeholders.
- Auditing and Monitoring: Establishing robust auditing and monitoring mechanisms is critical for maintaining visibility into data access and usage patterns within cloud BI environments. Regular audits, intrusion detection systems, and real-time monitoring tools enable organizations to detect suspicious activities, mitigate security incidents, and demonstrate compliance with regulatory requirements.
- Data Breach Response: Despite preventive measures, data breaches may still occur, necessitating a swift and coordinated response to minimize the impact on affected individuals and mitigate reputational damage. Developing a comprehensive incident response plan, conducting breach simulations, and collaborating with relevant stakeholders are essential components of an effective data breach response strategy.
- Vendor Security and Compliance: Evaluate the security practices and ensure that whether the CSPs has gone through regular security audits in place. Examine their compliance certifications, such as ISO 27001 or SOC 2, to verify they meet field benchmarks.
- Integration Complexities in Cloud BI: Integrate seamlessly with existing systems and data sources while migrating to a cloud-based BI solution, considering the complexities involved in data integration, transformation, and synchronization. Address challenges related to disparate data formats, incompatible schemas, and

varying data quality standards across different sources. Employ robust integration frameworks and technologies to streamline data flows and ensure compatibility between cloud BI platforms and existing IT infrastructures.

9 Collaboration and Team Dynamics in Cloud-Based BI Environments

In the era of cloud-based Business Intelligence (BI), the collaborative potential within organizational teams reaches new heights. The seamless integration of cloud BI platforms transforms traditional team dynamics, fostering a culture of real-time collaboration and data-driven decision-making. This paradigm shift is marked by geographical flexibility, as teams spanning diverse locations come together effortlessly. Cloud BI not only empowers cross-functional collaboration but also promotes a collaborative decision-making approach. With shared dashboards, discussion forums, and integrated task tracking, teams navigate a user-friendly interface that enhances accessibility. The following points delve into the intricacies of collaboration and team dynamics in cloud BI environments, showcasing the myriad ways in which these platforms revolutionize teamwork.

- Geographical Flexibility:

Cloud BI allows team members from different geographical locations to collaborate seamlessly.

- Real-time Data Sharing:

Team members can share and access real-time data through cloud BI platforms, promoting instant collaboration.

- Cross-Functional Collaboration:

Cloud BI breaks down silos, enabling collaboration among cross-functional teams such as marketing, finance, and operations.

- Shared Dashboards and Reports:

Teams can create and share dashboards and reports, ensuring everyone has access to the same insights.

- Version Control and Audit Trails:

Cloud BI platforms often provide version control and audit trails, enhancing accountability in collaborative efforts.

- Discussion Forums and Chat Features:

Many cloud BI tools offer discussion forums or chat features to facilitate communication and idea exchange among team members.

- Access Control and Permissions:

Cloud BI platforms allow fine-grained control over access permissions, ensuring data security while promoting collaboration.

- Task Assignment and Tracking:

Teams can assign tasks and track progress within cloud BI platforms, fostering a more organized and collaborative workflow.

- Collaborative Decision-Making:

Cloud BI enables teams to collectively analyze data, leading to more informed and collaborative decision-making processes.

- Notification Systems:

Users can set up notifications for changes or updates in data, ensuring that everyone stays informed and engaged.

- Integration with Collaboration Tools:

Integration with tools like Microsoft Teams or Slack facilitates seamless communication alongside data analysis.

- Training and Skill Development:

Collaboration in a cloud BI environment allows for skill-sharing and cross-training among team members.

- Data Ownership and Responsibility:

Clearly defined roles and responsibilities regarding data ownership contribute to a more collaborative environment.

- User-friendly Interfaces:

Intuitive interfaces in cloud BI tools make it easier for team members to collaborate without extensive training.

- Customizable Workspaces:

Cloud BI platforms often offer customizable workspaces to tailor the environment to the specific needs of different teams.

- Feedback Mechanisms:

Feedback features promote a culture of continuous improvement within the team.

- Collaborative Forecasting:

Teams can collaboratively work on forecasting models and predictions using cloud BI tools.

- Data Annotation and Documentation:

Cloud BI environments support data annotation and documentation, aiding in knowledge sharing and transfer.

- Project Management Integration:

Integration with project management tools streamlines collaboration by connecting data analysis with project workflows.

- Cross-Device Accessibility:

Team members can access and collaborate on cloud BI platforms from various devices, enhancing flexibility in work practices.

10 Scalability and Performance Optimization in Cloud BI

In the realm of Cloud Business Intelligence (BI), the twin imperatives of scalability and performance optimization stand as pillars, defining the efficacy of data-driven decision-making. As organizations grapple with ever-expanding datasets and dynamic workloads, the scalability of BI systems becomes paramount. Cloud BI platforms, equipped with innovative technologies, offer a transformative approach by seamlessly adapting to fluctuating demands. This adaptability is achieved through dynamic resource allocation, elastic scaling, and automated provisioning, laying the foundation for efficient performance optimization. In the following points, we delve into the intricacies of how cloud BI platforms ensure scalability and optimize performance, exploring the arsenal of strategies from parallel processing to predictive scaling that underpin the responsiveness and agility of modern BI environments.

- Dynamic Resource Allocation:

Cloud BI platforms dynamically allocate resources based on demand, ensuring scalability during peak usage.

- Elastic Scaling for Workloads:

Cloud BI allows for elastic scaling, enabling organizations to seamlessly adapt to changing workloads.

- Automated Provisioning of Resources:

The automation of resource provisioning in cloud BI ensures efficient scalability without manual intervention.

- Parallel Processing Capabilities:

Cloud BI platforms often leverage parallel processing to enhance performance and scalability for large datasets.

- Horizontal Scaling Options:

Organizations can achieve scalability by horizontally scaling resources, distributing the workload across multiple servers.

- Load Balancing Strategies:

Cloud BI environments employ load balancing techniques to evenly distribute workloads, optimizing performance.

- Caching Mechanisms for Frequently Accessed Data:

Implementing caching mechanisms for frequently accessed data enhances performance by reducing query response times.

- In-Memory Computing Technologies:

Cloud BI platforms leverage in-memory computing technologies to speed up data processing and analysis.

- Optimized Query Execution Plans:

Cloud BI tools generate optimized query execution plans to improve the efficiency of data retrieval.

- Data Partitioning Strategies:

Partitioning large datasets strategically facilitates efficient data retrieval and contributes to scalability.

- Compression Techniques for Data Storage:

Utilizing compression techniques in cloud BI environments minimizes storage requirements and boosts performance.

- Indexing Strategies for Quick Data Retrieval:

Effective indexing enhances the speed of data retrieval, optimizing the performance of cloud BI queries.

- Data Archiving and Tiered Storage:

Implementing data archiving and tiered storage ensures optimal use of resources and scalability for historical data.

- Predictive Scaling Algorithms:

Cloud BI platforms often employ predictive scaling algorithms to anticipate demand and scale resources proactively.

- Intelligent Caching for Dashboard Rendering:

Intelligent caching mechanisms optimize the rendering of dashboards, improving overall user experience.

- Fine-Tuned Query Optimization:

Continuous fine-tuning of query optimization algorithms in cloud BI platforms ensures sustained performance improvements.

- Optimized Network Bandwidth Usage:

Efficient use of network bandwidth reduces latency and contributes to the overall scalability of cloud BI systems.

- Auto-Scaling Policies Based on Performance Metrics:

Auto-scaling policies can be defined based on performance metrics, allowing resources to scale dynamically.

- Data Sharding for Parallel Processing:

Cloud BI systems leverage data sharding to distribute data across multiple nodes, enabling parallel processing.

- Continuous Monitoring and Performance Tuning:

Regular monitoring and performance tuning in cloud BI environments ensure ongoing scalability and optimal system performance [21].

11 Future Trends and Emerging Technologies in Cloud-Based Business Intelligence

In the rapidly evolving landscape of cloud-based Business Intelligence (BI), staying ahead necessitates a keen awareness of emerging technologies, evolving best practices, and anticipated shifts in industry trends. As organizations increasingly leverage the convergence of cloud computing and BI, the horizon is aglow with possibilities. From harnessing the power of Artificial Intelligence (AI) and Machine Learning (ML) to embedding analytics directly into business workflows, the journey toward enhanced decision-making and productivity is paved with innovation. Edge analytics, augmented analytics, and the democratization of data further propel this trajectory, promising real-time insights and seamless integration across diverse cloud environments. Moreover, as hybrid and multi-cloud deployments become the norm, advanced visualization techniques and robust data governance structures are set to redefine the paradigm of data-driven decision-making. In this section, we delve into the transformative potential of these trends and technologies, offering a glimpse into the future of cloud-based BI and its profound impact on organizational agility and competitiveness.

1. Artificial Intelligence (AI) and Machine Learning (ML): AI and ML technologies are playing a important roles in enhancing cloud BI. These technologies provide complex data analytics, AI-driven intelligence, and prognostic abilities, enabling enterprises to extract valuable insights from Big Data. Non-Technical users can

easily access cloud BI with the help of Natural Language Processing (NLP) powered by AI [22].

2. Embedded Analytics: Embedded analytics allows enterprises to provide real-time valuable insights and enhanced decision-making capabilities by directly integrating cloud BI into businesses. Embedded analytics is becoming more widespread in terms of use in cloud BI as it enables smooth integration and real-time analytics within different business work stream.
3. Edge Analytics: Edge analytics offers processing and analysis of data at or near the edge devices due to the emergence of IOT. It offers reduction in response time, improved real-time data-driven decision-making, and enhanced bandwidth usage [23].
4. Augmented Analytics: Augmented analytics harnesses AI and ML to mechanize data pre-processing, valuable insights generation, and suggestions within the BI work stream. It assists users with the help of Natural Language Processing (NLP). Augmented analytics is becoming a very significant trend in Cloud BI as it provides non-technical users to leverage cloud BI very easily.
5. Data Democratization: Data democratization targets to make data and analytics available to a more extensive range of users within the enterprise. Cloud BI solutions enable data democratization by offering self-service analytics tools and user-friendly interfaces or intuitive UI/UX. This fad enables business users to explore data, generate reports, and obtain valuable insights without depending heavily upon IT or data experts.
6. Hybrid and Multi-Cloud Deployments: Enterprises are progressively embracing hybrid and multi-cloud strategies to harness the power of different cloud service providers and sustain control over their data and applications or implementations. This trend effects cloud BI by ensuring effortless integration and data transfer between various cloud platforms, enabling data uniformity, and providing flexibility in terms of framework selections.
7. Advanced Visualization and Data Storytelling: Visualizations and data storytelling methods or tactics are acquiring eminence in cloud-based BI solutions. Sophisticated data visual presentation features ensure interactive and immersive data exploration/data investigation, enabling users to discover valuable insights and patterns more efficiently. Data storytelling methodologies help communicate valuable insights in a captivating and comprehensible manner, ensuring decision-makers to make informed choices in accordance with data-driven narratives.
8. Data Governance and Ethical AI: As enterprises depend more on cloud-based BI solutions and AI advancements, data stewardship and ethical considerations become crucial. The future of cloud-based BI solutions will witness a greater accentuation on establishing robust data governance infrastructures, guaranteeing data privacy, and addressing ethical concerns or considerations related to artificial intelligence algorithms and decision-making processes [24].

12 Ethical and Legal Implications of Cloud Computing in BI

In the rapidly evolving landscape of cloud-based Business Intelligence (BI), ethical and legal considerations loom large. As organizations increasingly rely on cloud computing to drive BI initiatives, questions surrounding data privacy, security, and regulatory compliance become paramount. In this section, we explore the ethical and legal implications of cloud computing in BI, shedding light on the challenges and responsibilities facing organizations in this digital age.

Ethical Implications

Data Privacy and Security

- Ensuring the protection of sensitive information in the cloud.
- Guaranteeing transparency, openness, and informed consent.

Bias in Data Analysis and Decision-Making

- Monitoring algorithms and machine learning to promote fairness and equality.

Legal Implications

Compliance with Data Protection Regulations

- Conforming to government regulations such as GDPR when storing and processing personal data.

Data Autonomy and Sovereignty

- Understanding and adhering to laws regulating the location of cloud service providers.
- Compliance with data autonomy laws across international borders.

In conclusion, the ethical and legal consequences of cloud computing in BI includes data privacy or data confidentiality, security, biasness, legal adherence, and data autonomy. Prioritizing and emphasizing transparency, consent, fairness, and regulatory compliance is crucial for responsible or accountable implementation of cloud BI [25].

13 Conclusion

The integration of cloud computing and business intelligence, collectively known as Cloud BI, possesses huge potential for companies seeking to harness valuable insights for data-driven decision-making. They provide great scalability, cost-efficiency, enhanced decision-making, global accessibility and real-time decision making with data renewed every minute. Cloud BI has transformative potential. Cloud BI has various practical implementations in Big Data, Real-Time Analytics, Customer Relationship Management (CRM), E-Commerce, Education, data integration and ETL. In the economic front Cloud BI has a tremendous impact, it provides enhanced Business Agility, competitive edge, better Return on Investment (ROI), Increased operational efficiency, Improved performance monitoring and cutting edge mobility. It also enhances data visualization which helps the users to understand complex data very easily. It offers great data security measures that attracts more and more user to harness Cloud BI which in turn improves these organizations in almost every aspect, thus indirectly benefiting our economy. In conclusion, Cloud BI stands as a transformative force and a catalyst, driving businesses towards unparalleled growth and unmatched insights or extraordinary intelligence.

References

1. Ouf, S., Nasr, M.: Business intelligence in the cloud. In: 2011 IEEE 3rd International Conference on Communication Software and Networks, Xi'an, China, pp. 650–655 (2011). https://doi.org/10.1109/ICCSN.2011.6014351. https://ieeexplore.ieee.org/document/6014351
2. Wang, F.., Yan, A., Li, X., Yang, L.: Cloud computing based business intelligence platform and its application in the field of intelligent power consumption. In: 2011 International Conference on Electronics, Communications and Control (ICECC), Ningbo, China, pp. 3612–3616 (2011). https://doi.org/10.1109/ICECC.2011.6066291. https://ieeexplore.ieee.org/document/6066291
3. Ouf, S., Nasr, M.: Business intelligence in the cloud. In: 2011 IEEE 3rd International Conference on Communication Software and Networks, Xi'an, China, pp. 650–655 (2011). https://doi.org/10.1109/ICCSN.2011.6014351. https://ieeexplore.ieee.org/document/6014351?arnumber=6014351
4. Herwig,V.: Business intelligence as a service for cloud-based applications. In: 2013 IEEE 7th International Conference on Intelligent Data Acquisition and Advanced Computing Systems (IDAACS), Berlin, Germany, pp. 188–192 (2013). https://doi.org/10.1109/IDAACS.2013.6662668. https://ieeexplore.ieee.org/document/6662668
5. Zhang, J.: Research on the application of cloud computing technology in computer data processing. In: 2021 3rd International Conference on Machine Learning, Big Data and Business Intelligence (MLBDBI), Taiyuan, China, pp. 510–514 (2021). https://doi.org/10.1109/MLBDBI54094.2021.00103. https://ieeexplore.ieee.org/document/9731074
6. Khan, S., Zhang, B., Khan, F., Chen, S.: Business intelligence in the cloud: a case of Pakistan. In: 2011 IEEE International Conference on Cloud Computing and Intelligence Systems, Beijing, China, pp. 536–540 (2011). https://doi.org/10.1109/CCIS.2011.6045126. https://ieeexplore.ieee.org/document/6045126
7. Juan-Verdejo, A., Surajbali, B., Baars, H., Kemper, H.-G.: Moving business intelligence to cloud environments. In: 2014 IEEE Conference on Computer Communications Workshops (INFOCOM WKSHPS), Toronto, ON, Canada, pp. 43–48 (2014). https://doi.org/10.1109/INFCOMW.2014.6849166. https://ieeexplore.ieee.org/document/6849166
8. Alsufyani, R., Chang, V.: Risk analysis of business intelligence in cloud computing. In: 2015 IEEE 7th International Conference on Cloud Computing Technology and Science (CloudCom), Vancouver, BC, Canada, pp. 558–563 (2015). https://doi.org/10.1109/CloudCom.2015.84. https://ieeexplore.ieee.org/document/7396213
9. Gatzioura, A., Menychtas, A., Moulos, V., Varvarigou, T.: Incorporating business intelligence in cloud marketplaces. In: 2012 IEEE 10th International Symposium on Parallel and Distributed Processing with Applications, Leganes, Spain, pp. 466–472 (2012). https://doi.org/10.1109/ISPA.2012.68. https://ieeexplore.ieee.org/document/6280328

10. Pondel, M.: Business Intelligence as a service in a cloud environment. In: 2013 Federated Conference on Computer Science and Information Systems, Krakow, Poland, pp. 1281–1283. https://ieeexplore.ieee.org/abstract/document/6644179
11. Stipić, A., Bronzin, T.: How cloud computing is (not) changing the way we do BI. In: 2012 Proceedings of the 35th International Convention MIPRO, Opatija, Croatia, pp. 1574–1582 (2012). https://ieeexplore.ieee.org/document/6240903
12. Zhang, Y., Ma, H., Xu, Y.: An intelligence gathering system for business based on cloud computing. In: 2013 Sixth International Symposium on Computational Intelligence and Design, Hangzhou, China, pp. 201–204 (2013). https://doi.org/10.1109/ISCID.2013.57. https://ieeexplore.ieee.org/document/6804970
13. Al-Aqrabi, H., Liu, L., Hill, R., Cui, L., Li, J.: Faceted search in business intelligence on the cloud. In: 2013 IEEE International Conference on Green Computing and Communications and IEEE Internet of Things and IEEE Cyber, Physical and Social Computing, Beijing, China, pp. 842–849 (2013). https://doi.org/10.1109/GreenCom-iThings-CPSCom.2013.148. https://ieeexplore.ieee.org/document/6682162
14. Uchida, N., Takahata, K., Shibata, Y.: Proposal of business intelligence system by overlay cloud computing architecture. In: 2013 Eighth International Conference on P2P, Parallel, Grid, Cloud and Internet Computing, Compiegne, France, pp. 578–581 (2013). https://doi.org/10.1109/3PGCIC.2013.100. https://ieeexplore.ieee.org/document/6681294
15. Tole, A.A.: Cloud computing and business intelligence. Database Syst. J. (1970, January 1)https://ideas.repec.org/a/aes/dbjour/v5y2015i4p49-58.html
16. Thompson, W.J., Van der Walt, J.S.: Business intelligence in the cloud. South Afr. J. Inform. Manage. **12**(1), 1–15 (2010). https://doi.org/10.4102/sajim.v12i1.445
17. Mircea, M., Ghilic-Micu, B., Stoica, M.: Combining business intelligence with cloud computing to delivery agility in actual economy. ResearchGate (2011). https://www.researchgate.net/publication/290547197_Combining_business_intelligence_with_cloud_computing_to_delivery_agility_in_actual_economy
18. Al-Aqrabi, H., Liu, L., Hill, R., Antonopoulos, N.: Cloud BI: future of business intelligence in the cloud (2019). https://www.researchgate.net/publication/330617510_Cloud_BI_Future_of_Business_Intelligence_in_the_Cloud
19. Ouf, S., et al.: Business intelligence software as a service (SAAS). In: 2011 IEEE 3rd International Conference on Communication Software and Networks, Xi'an, China, pp. 641–649 (2011). https://doi.org/10.1109/ICCSN.2011.6014350. https://ieeexplore.ieee.org/document/6014350
20. Shahata, N.: How cloud computing and business intelligence support UAV technology. J. Network. Network Appl. **2**(3), 116–119 (2022). https://doi.org/10.33969/J-NaNA.2022.020303
21. Khouibiri, N., Farhaoui, Y.: Analyzing the influence of cloud business intelligence on small and medium enterprises a case study of Morocco. Data Metadata **2**, 104. Available from: https://dm.saludcyt.ar/index.php/dm/article/view/104
22. Saxena, A., Chauhan, R., Chauhan, D., Sharma, S., Sharma, D.: Comparative analysis of AI regression and classification models for predicting earthquake-induced house damages in Nepal. NeuroQuantology **20**(22), 1987–1999 (2022). https://doi.org/10.48047/nq.2022.20.22.NQ10186. https://www.pnrjournal.com/index.php/home/article/view/10497/14693
23. Soni, M., Nayak, N.R., Kalra, A., Degadwala, S., Singh, N.K., Singh, S.: Energy efficient multi-tasking for edge computing using federated learning. Int. J. Pervasive Comput. Commun. (2022). https://doi.org/10.1108/IJPCC-03-2022-0106
24. Kumar, V., Himanshu, Rawat, A., Natarajan, S.K.: A deep dive on business intelligence systems and infrastructure using cloud environment. 1–5 (2021). https://doi.org/10.1109/INCET51464.2021.9456369. https://www.researchgate.net/publication/352671947_A_Deep_Dive_on_Business_Intelligence_Systems_and_Infrastructure_using_Cloud_Environment
25. Sultan, H., Chan, L., Siew, P., Lee, C., Cheah, C., Ong, J.: Cloud business intelligence issues and challenges. Int. J. Technol. Manage. Inform. Syst. **4**(4), 75–85 (2022). Retrieved from https://myjms.mohe.gov.my/index.php/ijtmis/article/view/20852

Predicting Student Performance Through Demographic Attributes

Abhishek Kumar

Abstract Education is essential in determining how a student's life will unfold. There are a lot of alternatives available to students when selecting a career path. As the cornerstone of the classification training system, factors like talents, abilities, family background, and type of schooling become critical in directing pupils toward a certain career track. A linear regression model is utilized in this chapter to clarify the complex relationships that exist between important demographic variables and academic achievement. This research adds to a thorough knowledge of the complex variables affecting students' achievement and provides information for well-informed interventions and support techniques in the context of education. The study also explores how machine learning may improve the educational system.

Keywords Student performance prediction · Linear regression algorithm · Academic assessment · Attendance and evaluation · Final examination forecast · Machine learning in education · Personal capacity in academic success · Gender disparities

1 Introduction

For educators and politicians, understanding the variables influencing students' performance is crucial in the ever-changing field of education [1]. This chapter explores how to use a Linear Regression Model to predict student performance by utilizing important independent factors including test preparation, lunch, gender, race/ethnicity, and education level. Predicting performance in three critical areas—math, reading, and writing—is the main focus. Accomplishment understanding and prediction are critical in the ever-changing field of education. By anticipating academic results, teachers can better customize interventions, giving schools

A. Kumar (✉)
School of Electronics and Electrical Engineering, Lovely Professional University, Jalandhar, Punjab, India
e-mail: abkvjti@gmail.com

S. Mahajan et al. (eds.), *Smart Systems: Engineering and Managing Information for Future Success*, Information Systems Engineering and Management 22,
https://doi.org/10.1007/978-3-031-76152-2_13

the flexibility to create curricula that are tailored to the varied needs of their student body [2]. This chapter delves deeply into the complex relationship between student demographics and academic success, highlighting important aspects of this relationship.

While selecting the career path, a student has multiple alternatives like traditional education, vocational training, coaching models, apprenticeships, and community college programs. Additionally, it encompasses the spectrum of career choices available to individuals post-education, including entry-level positions, internships, industry-specific roles, entrepreneurship, freelance opportunities, and advanced professional pathways. Examining the choices available to students, the scope attempts to provide an in-depth understanding of the variables impacting career decisions and the range of options that students may select from to accomplish their professional objectives. It takes into account factors including academic interests, skill development, industry developments, demands of the labor market, financial concerns, and personal goals. Predictive modeling of student performance offers invaluable insights into the factors influencing academic success outside of the typical academic setting. By examining the complex web of student performance, teachers can proactively identify barriers, implement customized learning strategies, and establish a supportive environment for academic success [3]. Predictive analytics can also help companies better manage resources, which will enhance the educational experiences of students. Demographic factors such as race, gender, test-taking strategies, and lunchtime have a big impact on how well children perform academically. These factors capture the nuanced facets of diversity and socioeconomic origins, which have an impact on educational opportunities and outcomes. Gaining an understanding of the importance of these traits is necessary to create inclusive and fair learning environments and ensure that each student receives customized assistance. In light of the above considerations, this study seeks to address the following research question examines academic performance.

How do demographic attributes, including Gender, Race, Lunch, and Test Preparation, correlate with and impact student performance in academic subjects, specifically Math, Reading, and Writing scores?

The objective of the work is

(a) *To analyze the influence of Gender on academic performance:* examination of any gender-based trends or disparities in Math, Reading, and Writing scores.
(b) *To investigate the relationship between Race and academic outcomes*: Identifying potential cultural or socio-economic factors contributing to differences in performance.
(c) *To examine the effect of Lunch on academic scores*: potential implications related to socio-economic status.
(d) *To evaluate the correlation between Test Preparation and academic performance*: analyzing the significance of preparation strategies on Math, Reading, and Writing scores.

(e) *To explore the integrated impact of multiple demographic attributes*: Investigating how the combination of Gender, Race, Lunch, and Test Preparation influences overall academic performance

2 Literature Review

In the field of education, a great deal of study has been conducted to better understand the variables that affect student performance. The purpose of this review of the literature is to compile the research that has already been done on the prediction of student results, with a focus on the effects of demographic characteristics including test preparation, lunch, gender, and race [4]. Predicting student performance has been the subject of research using a variety of approaches and investigating a wide range of criteria. A collection of studies has examined the ability of demographic characteristics to predict outcomes. Smith et al. [5] investigated the long-term effects of gender on academic attainment using longitudinal data; their findings are male and female students differ slightly in their preferred subjects and learning styles, which has an impact on their academic performance. Thompson and Carter in [6], concentrated on how socioeconomic issues, including lunch programs, affect people's lives. Their study demonstrated how socioeconomic status shapes academic accomplishment, with students who get free or reduced lunches encountering extra obstacles that have an impact on their general performance. studies conducted in [7] on test preparation techniques revealed a favorable correlation between deliberate preparation and better academic performance. Even while previous research has provided insightful information, there are still some significant gaps that need to be filled.

A student's life is greatly shaped by their education, which provides a wide range of job alternatives. During the decision-making process, students take into account their talents, abilities, family history, and educational background—all of which are important determinants that help them choose a career path. These elements also play a part in the system of classification training preparation, and investigate the use of machine learning techniques in remote prediction, making it a trailblazing work [8]. Their findings served as a foundation for later studies, including our own. Although machine learning (ML) techniques are widely used, their incorporation into scientific contexts is still a relatively recent development. benefits of a variety of unique characteristics, or perhaps aspects, is acknowledged to confirm which of them are connected with student abilities The goal of a study by [9] was to predict students' performance. Machine learning uses incremental learning technology to handle new features and data while retaining previously taught knowledge. The method known as incremental learning makes use of past data to make decisions and change the program going forward. It is expected that a student will perform well on the final exam if they regularly attend classes, perform well on class tests, and do well on evaluations [10]. Nonetheless, situations can occur when a student does well on the midterm but has trouble showing up to class or with examinations. On the other hand, a student could have outstanding presentation abilities but struggle in the first exam. While we concede that perfection is impossible, particularly for students, we maintain that we are all capable of performing to the best of our abilities. Using past event reports from students, this study aims to predict final exam results. With the use of machine learning and the linear regression technique, the study attempts to forecast future outcomes based on current qualities.

Researchers have shown an extensive interest in predicting students' academic performance, looking at a variety of elements that may lead to success or difficulties in learning environments. An important line of inquiry is to look at demographic characteristics as possible predictors. The important conclusions from pertinent studies are compiled in this overview of the literature, illuminating the complex relationship between student performance and demographic characteristics. Understanding the impact of demographic attributes on predicting student performance is crucial, necessitating a nuanced approach to this complex topic. The multifaceted nature of educational outcomes underscores the need for comprehensive exploration. Future research should persist in investigating these relationships, taking into account the dynamic nature of demographics and the evolving educational landscape. Recognizing and addressing the various factors influencing student performance will be instrumental in shaping targeted interventions and strategies geared toward fostering academic success. This approach acknowledges the diverse and evolving nature of student populations, paving the way for tailored solutions that cater to the specific challenges and strengths associated with different demographic attributes.

3 Majorly Researched Demographic Attributes Are as Follows

(i) ***Social and Economic Status***

Studies by [11] and [12] highlight how socioeconomic status affects academic performance. It has been discovered that variables including parental education, income, and occupation affect student accomplishment and frequently act as predisposing factors for differences in academic success.

(ii) ***Gender Disparities***

Studies on gender variations in academic achievement can be found continuing, as demonstrated by the work of [13] and [14]. Even while some research points to gender differences in particular disciplines, the overall effect of gender on student performance is still a complicated topic that needs more research.

(iii) ***Cultural and Ethnic Influences***

Researchers in [15] and [16] have investigated the significance of cultural and ethnic background in determining student achievement. These studies highlight how crucial it is to take cultural context into account when assessing academic accomplishment.

(iv) ***Language Proficiency***

Academic achievement is significantly influenced by language competency, especially for students from non-native language backgrounds. Researchers in [17] and [18] examine the connection between language proficiency, cognitive growth, and academic achievement.

(v) ***Learning Styles and Preferences***

The importance of taking into account each learner's unique preferences and learning styles is demonstrated by the works of [19] and [20]. Personalized educational strategies that improve overall performance can be informed by an understanding of the diverse ways that pupils receive information.

(vi) ***Parental Involvement***

Parental participation: Research by [21, 22] shows that parental participation is a key factor. Collaborations between educators, parents, and students have demonstrated beneficial relationships with academic success.

(vii) ***Technology Access and Usage***

Technology is essential in today's educational environment. Research on how technology use and access affect student performance has been conducted by [23] and [24]. These studies show a complex relationship between digital resources and academic results.

4 Data Collection and Preprocessing

The dataset is meticulously collected from [25], comprising information on Gender, Race/Ethnicity, Level of Education, Lunch status, Test Preparation, and academic scores in Math, Reading, and Writing. Data preprocessing techniques are applied to handle missing values, and outliers, and ensure the dataset's readiness for modeling.

Descriptive statistics were used to thoroughly analyze the dataset and provide important new information on the demographic characteristics and academic standing of the students presented in Table 1. The gender distribution showed a balanced representation, with the percentage of male and female pupils shown graphically. An examination of racial or ethnic backgrounds revealed a varied student body, with bar charts clarifying the prevalence among various groupings [27]. The socioeconomic environment was outlined by lunch program participation Figs, which captured the choice between standard and free/reduced meal options. Descriptive statistics and visual aids were also utilized to analyze the degree of participation in test-taking programs. A thorough analysis of the academic results in math, reading, and writing was conducted. The distribution curve of student performance factors and various attributes contribute to the complex tapestry of academic outcomes. The distribution curve in Fig. 1 presents reading and writing scores are distributed 40–80. Few

Table 1 Attribute affecting student performance [26]

Attribute	Description
Gender	Male or female
Race/ethnicity	Group A/B/C/D/E
Lunch	Standard or free/reduced
Test preparation	None or completed

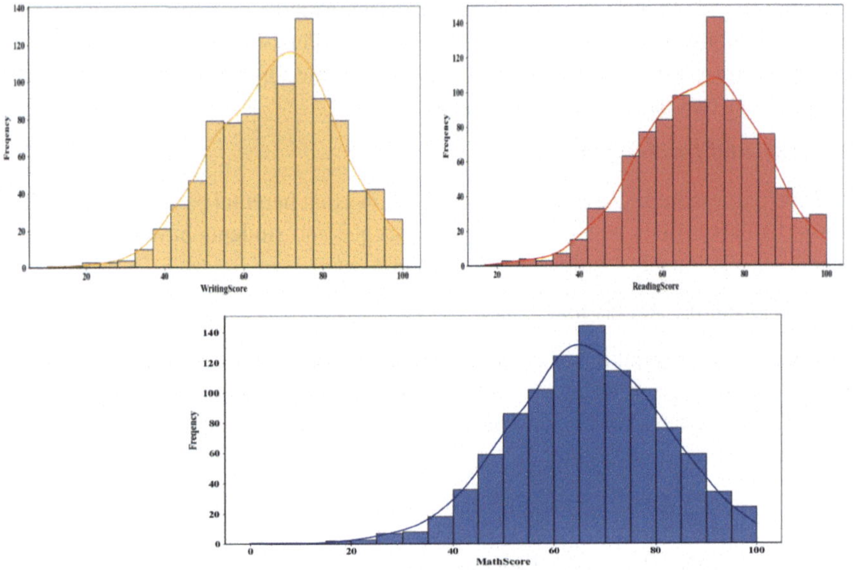

Fig. 1 Distribution curves of writing, reading, and math score

student's scores peak. While in math score distribution curve incremental distributed lossless number of outliers.

5 Methodology

This research aims to categorize and identify the future talents of students by analyzing school records and considering a significant influencing factor. This process is crucial for implementing effective strategies and pinpointing areas where students may need additional support. The outcomes of this study will not only contribute to curriculum and educational policy development but also inform placement strategies. In essence, the research focuses on employing various standard machine-learning classification models on collected datasets and comparing them with a holistic machine-learning approach that utilizes data mining tools. This research is based on an extensive dataset that includes a variety of student characteristics and measures of academic achievement. The dataset includes data on variables like gender, race, lunch participation, test preparation, and math, reading, and writing academic achievements shown in Fig. 2. Testing programs and educational institutions are among the data sources, guaranteeing a large and representative sample [28]. Pre-processing steps ensure the dataset is comprehensive, covering a diverse range of student profiles. It handles missing data and removes outliers appropriately. Encoding of categorical variables using techniques like one-hot encoding and normalizing to numerical features to bring them to a consistent scale. Split the dataset into training and testing sets at the ratio of 80% for model evaluation.

5.1 Variables

(a) Gender: Categorical variable representing the gender of each student (Male, Female).
(b) Race: Categorical variable indicating the racial or ethnic background of each student (Group A, Group B, Group C, Group D, Group E).
(c) Lunch: Categorical variable denoting lunch program participation, serving as a proxy for socio-economic status (Standard, Free/Reduced).
(d) Test Preparation: Categorical variable indicating whether a student completed a test preparation course (Completed, None).
(e) Math Score: Continuous variable representing the academic score in Mathematics.
(f) Reading Score: Continuous variable representing the academic score in Reading.
(g) Writing Score: Continuous variable representing the academic score in Writing.

The statistical technique used in this study to predict academic outcomes based on demographic characteristics is linear regression. Using this approach, one can investigate linear correlations between dependent variables (math, reading, and writing scores) and independent variables (gender, race, lunch, and test preparation). To make predictions for new or unknown data points, the linear regression model seeks to find the mathematical formula of Eq. (1) that most closely matches the observed data [14].

$$Y_i = \beta_0 + \beta_1 X_1 + \beta_2 X_2 + \beta_3 X_3 + \beta_4 X_4 + \varepsilon_i \quad (1)$$

where Y_i = predicted score, β_0 is the intercept, $\beta_1 - \beta_4$ is the coefficient with independent variables, $X_1 - X_4$ are values of Gender, Race, Lunch, and Test prediction for the particular student, and ε is the error.

Fig. 2 Flow chart of the methodology

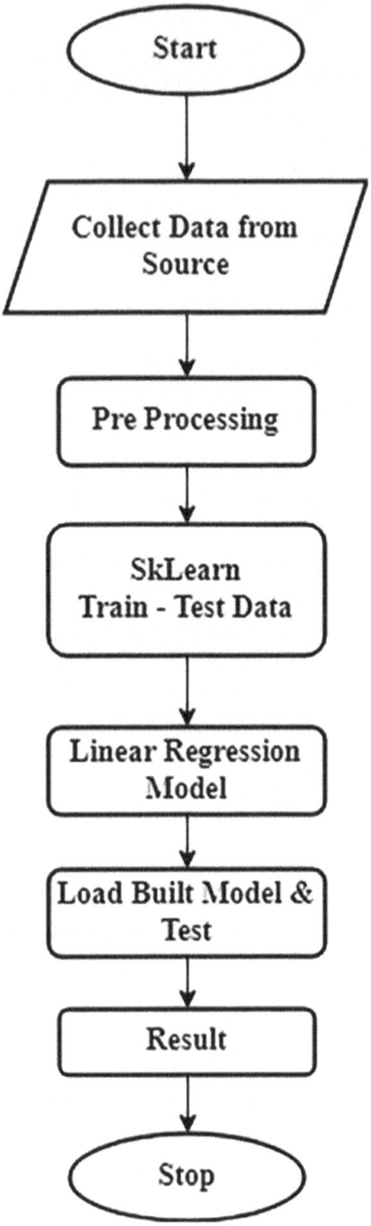

5.2 Linear Regression Model

Machine learning holds significant potential for revolutionizing the educational system by offering personalized interventions, optimizing teaching methodologies, and improving learning outcomes. machine learning algorithms can evaluate student data to create individualized learning routes for each student, such as performance indicators, learning preferences, and styles. Adaptive learning systems can modify the degree of challenge in exercises, suggest pertinent resources, and offer focused feedback according to the individual strengths and limitations of each student. To predict future academic outcomes and trends, machine learning algorithms can examine databases containing student performance, demographic data, and socioeconomic characteristics. To increase student success rates, educational institutions can use these predictive analytics to pinpoint systemic issues, manage resources wisely, and carry out focused interventions. The cornerstone of this predictive analysis is the Linear Regression Model. A comprehensive overview of the model is provided, emphasizing its suitability for predicting continuous dependent variables. The linear relationship between the independent variables and academic scores is explored, with discussions on assumptions, model fitting, and interpretation of coefficients. Preparing the data is an essential stage in ensuring the accuracy and dependability of the outcomes. There is a requirement of the SKLEARN library for implementing the Linear Regression model. A Linear Regression model to establish relationships between independent variables (Gender, Race/Ethnicity, Level of Education, Lunch, Test Preparation) and dependent variables (Math, Reading, and Writing scores). Train the model using the training dataset, allowing it to learn patterns and correlations.

(a) Handling Missing Data: To ensure that the dataset is complete, any missing values have been properly addressed through imputation or removal.
(b) Coding Categorical Variables: Using methods like one-hot encoding, categorical variables such as gender, race, test preparation, and lunch have been converted into numerical formats so they may be included in the regression model. Race/ethnicity comprises 5 groups (A/B/C/D/E) are encoded as 0–5. Encoding of luch performed as standard – 1 and free/reduced as 0. Similarly gender encoding male-1 and 0-female. test preparation encoding yes-1 no-0. Encoding for the level of education follows 0-high school, 1—bachelor's degree, 2—some college, 3—same high school, 4—master's degree, 5—associate's degree
(c) Feature Scaling: To avoid some variables from dominating due to their scale, continuous variables have been scaled to a common range.
(d) Data Splitting: To make model training and evaluation easier, the dataset has been divided into training and testing sets at 80% level.

The linear regression model investigates the relationships between different demographic characteristics and the results of academic performance, highlighting key variables. The diversity in math scores related to race is depicted in Fig. 3, which highlights potential differences between various racial or ethnic groupings. Interestingly, students who study well for exams have higher math scores, demonstrating

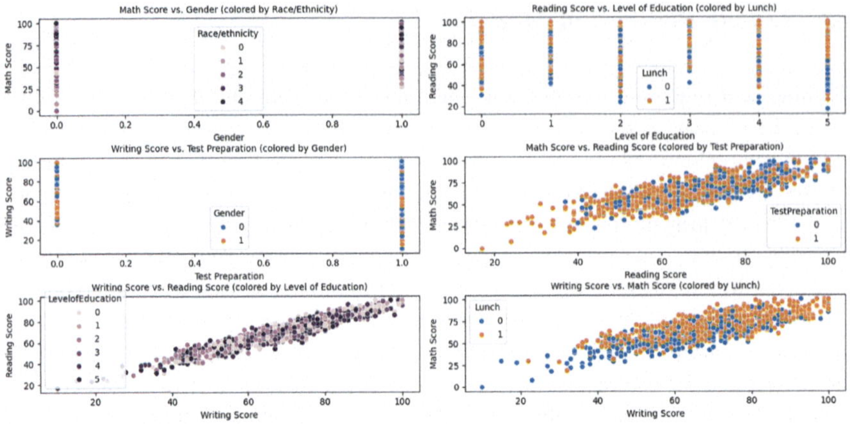

Fig. 3 Scatter plot of the variable

the beneficial effects of test preparation on academic achievement. Low-income students may be eligible for free or subsidized meals, and the lunch variable is a crucial indicator of socioeconomic status. Research indicates that pupils who regularly eat lunches tend to attain higher reading scores, emphasizing the relationship between academic success and social status. When it comes to writing scores, gender differences show up, suggesting that male students might be better writers even in a lack of test preparation. On the other hand, female students exhibit an even stronger association between high writing scores and preparation levels, indicating a complex interaction between academic outcomes, test preparation, and gender. Figure 3 illustrates how lunch affects math and writing scores. Regular lunch participants exhibit greater writing and math skills, highlighting the relationship between eating habits and scholastic achievement. This comprehensive examination of demographic characteristics advances our knowledge of the complex factors influencing academic success and directs resources and actions to provide fair educational opportunities for all pupils.

5.3 Scatter Plot

Examining the complex dynamics of student performance, a study was conducted to determine the influence of each variable on scores. Gender analysis provided detailed information about math, reading, and writing scores, highlighting any trends or discrepancies. The impact of different racial or ethnic backgrounds on educational results was revealed through a methodical investigation of the relationship between race and academic performance for each topic. A detailed analysis of the impact of participation in the Lunch Program on Math, Reading, and Writing scores shed light on the influence of socioeconomic factors on academic performance. Insights

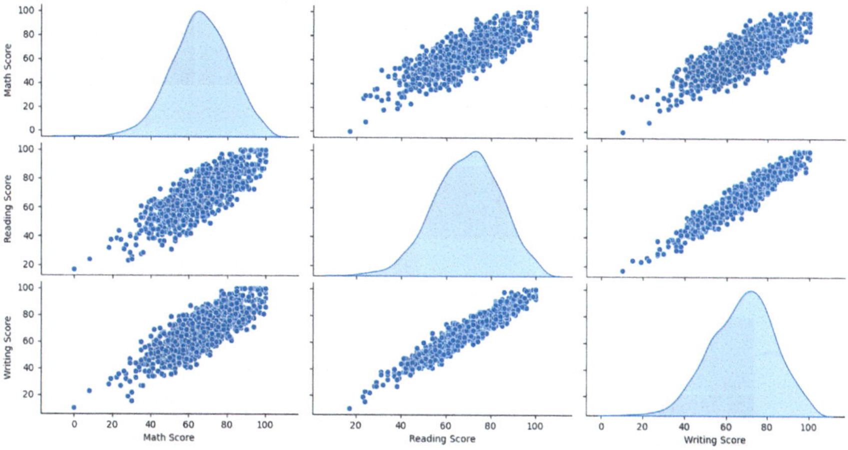

Fig. 4 Pair plot of variables

standards lunch required to perform well in examinations. The relationship between test preparation and performance in each subject was evaluated, offering a thorough grasp of how preparation affects test results. a comprehensive analysis was carried out to determine the overall effects of the combination of Gender, Race, Lunch, and Test Preparation on academic performance. Graph insights test preparation improves the writing score. This comprehensive study established in Fig. 3 is the foundation for a sophisticated understanding of the relationship between demographic characteristics and academic achievement, promoting well-informed tactics for academic assistance and intervention.

5.4 *Pair Plot*

A pair plot in Fig. 4 analysis is useful in the effort to comprehend the complex links that exist between different student qualities and academic achievement. Scatterplots for pairs of variables are visually represented by a pair plot, which provides information about possible relationships and patterns [29]. Reading and writing scores are less correlated compared with math score.

5.5 *Correlation Heatmap*

An effective technique for illustrating the direction and intensity of correlations between several variables is a correlation heatmap presented in Fig. 5. The heatmap

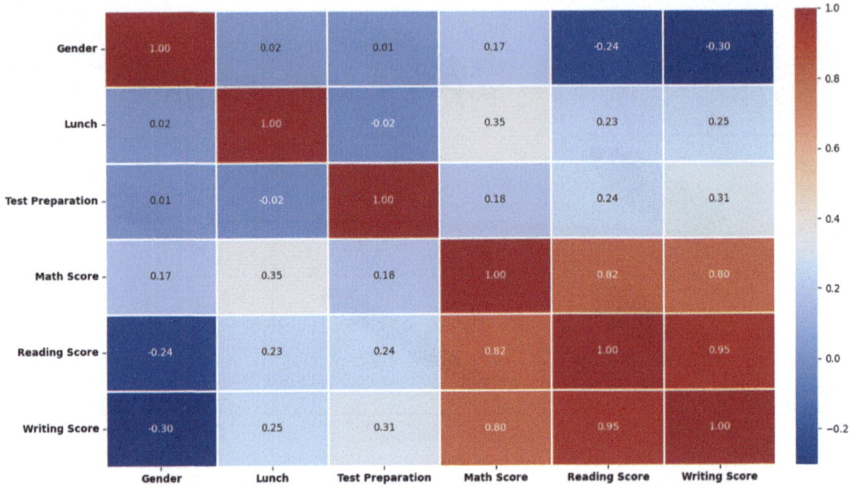

Fig. 5 Correlation heatmap of variables

offers a visual representation of correlations between important attributes in the context of predicting student performance [30, 31].

A negative correlation value between gender and reading/writing score presents an inverse relationship. A weak or negligible correlation is found between gender math scores, gender test preparation, and gender lunch. A moderate correlation between scores with lunch, and test preparation. A strong correlation presented in red color, insights the improvement in reading and writing scores significantly increases the math scores.

5.6 *Variable Importance*

Each independent variable's impact on student performance is assessed. The chapter elucidates the significance of Gender, Race/Ethnicity, Level of Education, Lunch, and Test Preparation in influencing Math, Reading, and Writing scores. Insights gained from variable importance help educators tailor interventions and support systems. Figure 6 shows that test preparation is the most important variable to improve scores followed by gender, lunch, race, and lastly level of education.

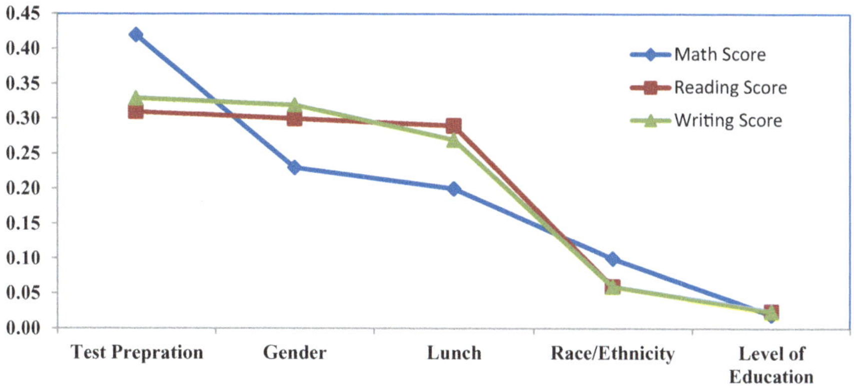

Fig. 6 Relative importance curve

6 Model Evaluation

The efficacy of the Linear Regression Model is rigorously evaluated [32]. Using a different dataset, the trained model is put through a rigorous testing process to determine its predictive power during the evaluation phase. This testing dataset provides a baseline against which to evaluate the model's resilience and determine how well it generalizes to new cases. The flexibility of the model and its capacity to generate precise predictions based on newly available data is demonstrated by adding new data points. Keeping reading and writing scores at 90% the predicted math score achieves 80.05%.

A thorough study of the results is carried out by contrasting the model's predictions with the actual results. Table 2 presents the model's performance, metrics like R-squared or Mean Squared Error (MSE) are used during the training and testing phase. These evaluations provide important insights into the accuracy of the model, facilitating a comprehensive comprehension of its predictive power and providing guidance for future improvements or optimizations. The practical implications of these metrics for educational decision-making are discussed.

During the training phase R^2 value 0.92501 indicates that approximately 92.5% of the variability in the response variable (target) is explained, while adjusted R^2 suggests model explanatory power is 92.43%, Variance Accounted For of 0.92501 aligns with the R^2 value. Root Mean Squared Error (RMSE) of 3.957 represents the average magnitude of the residuals, Mean Absolute Error (MAE) between actual

Table 2 Regression model evaluation

	R^2	Adjusted R^2	VAF	RMSE	MAE	PI	Rank
Training	0.92501	0.92435	0.92501	3.9577	3.1977	0.04208	– 4.3390
Testing	0.91525	0.91216	0.91526	4.3791	3.5109	0.04282	– 5.1046

and predicted value 3.1977. A prediction interval of 0.04208 suggests a relatively narrow and negative rank value implying that the model performs well relative to the benchmarks used for ranking.

During the Testing phase, An R^2 of 0.91525 indicates that approximately 91.5% of the variability in the response variable is explained by the model, however model has 91.216% of explanatory power indicated by adjusted R^2. The value of VAF 0.91526 aligns closely with the R^2 value, indicating the proportion of variance captured by the model. RMSE value of 4.3791 presents the prediction error with a mean absolute error is 3.51. The value of PI 0.04282 suggests a relatively narrow prediction interval, indicating a higher level of confidence in the model's predictions.

7 Educational Insights

The predictive model offers valuable insights for educators, administrators, and policymakers. It highlights potential areas for targeted interventions and sheds light on the nuanced interplay of various factors affecting student outcomes. Predict the student's academic performance as well as future prediction. They will be helpful for teachers, parents, and students. The teacher will provide proper guidance and direction, and then improve the performance of students and bad results. The target population for the results of these studies, which investigate the association between demographic characteristics and academic achievement through the use of a linear regression model, includes education researchers, policymakers, and educators.

Educators gain insights from knowing how demographic factors, including gender, race, test-taking strategies, and lunch habits impact their students' academic achievement. Policymakers in education can use these findings to inform policies aimed at addressing disparities in academic achievement and promoting equity in education. Researchers in the field of education can build upon these findings to conduct further investigations into the complex interactions between demographic variables and academic performance. They can explore additional factors, refine methodologies, and examine longitudinal trends to deepen understanding and inform evidence-based practices in education. For educators, policymakers, and researchers, the research findings aim to contribute to ongoing discussions and efforts to promote educational equity, improve student outcomes, and foster inclusive learning environments.

8 Conclusion

The chapter concludes by summarizing key findings, emphasizing the importance of utilizing predictive models in education. It calls for further research, collaboration, and the integration of data-driven insights to foster a more inclusive and supportive educational environment. A linear regression model predicts student performance is effective in explaining 91.5% of the variance in academic outcomes. Analyzing important demographic characteristics like gender, race/ethnicity, education level, lunch, and test preparation has provided important new insights into the intricate interactions between variables affecting academic success. The model's predictive ability (R^2 0.91525) highlights how useful it is for predicting student achievement. With the use of this predictive capacity, educators can focus on particular areas that have an impact on academic achievement and implement tailored interventions. The knowledge gathered from this research provides a road map for the ongoing development of instructional tactics. Through the utilization of machine learning models, academic institutions can customize their strategies, execute focused interventions, and cultivate an atmosphere that optimizes students' prospects for both academic achievement and overall growth.

References

1. Price, D., McCallum, F.: Ecological influences on teachers' well-being and "fitness." Asia-Pacific J. Teacher Educ. **43**(3), 195–209 (2015)
2. Linn, M.C., Clark, D., Slotta, J.D.: WISE design for knowledge integration. Sci. Educ. **87**(4), 517–538 (2003)
3. Namoun, A., Alshanqiti, A.: Predicting student performance using data mining and learning analytics techniques: a systematic literature review. Appl. Sci. **11**(1), 237 (2020)
4. Wiswall, M., Stiefel, L., Schwartz, A.E., Boccardo, J.: Does attending a STEM high school improve student performance? Evidence from New York City. Econ. Educ. Rev. **40**, 93–105 (2014)
5. Smith, S.L., Kindermans, P.J., Ying, C., Le, Q.V.: Don't decay the learning rate, increase the batch size (2017). arXiv preprint arXiv:1711.00489.
6. Thomson, K., Hillier-Brown, F., Todd, A., McNamara, C., Huijts, T., Bambra, C.: The effects of public health policies on health inequalities in high-income countries: an umbrella review. BMC Public Health **18**, 1–21 (2018)
7. Cassady, J.C., Johnson, R.E.: Cognitive test anxiety and academic performance. Contemp. Educ. Psychol. **27**(2), 270–295 (2002)
8. Koseler Emre, R.: Developing a neural network based adaptive task selection system for anundergraduate level organic chemistry course. Arizona State University (2020)
9. He, H., Chen, S., Li, K., Xu, X.: Incremental learning from stream data. IEEE Trans. Neural Networks **22**(12), 1901–1914 (2011)
10. Dobkin, C., Gil, R., Marion, J.: Skipping class in college and exam performance: evidence from a regression discontinuity classroom experiment. Econ. Educ. Rev. **29**(4), 566–575 (2010)
11. Reardon, S.F.: The widening academic achievement gap between the rich and the poor: new evidence and possible explanations. Whither Opportunity **1**(1), 91–116 (2011)
12. Sirin, S.R.: Socioeconomic status and academic achievement: a meta-analytic review of research. Rev. Educ. Res. **75**(3), 417–453 (2005)
13. Hyde, J.S., Lindberg, S.M., Linn, M.C., Ellis, A.B., Williams, C.C.: Gender similarities characterize math performance. Science **321**(5888), 494–495 (2008)
14. Voyer, D., Voyer, S.D.: Gender differences in scholastic achievement: a meta-analysis. Psychol. Bull. **140**(4), 1174 (2014)
15. Oyserman, D., Destin, M.: Identity-based motivation: implications for intervention. Couns. Psychol. **38**(7), 1001–1043 (2010)
16. Ream, R.K., Palardy, G.J.: Reexamining social class differences in the availability and the educational utility of parental social capital. Am. Educ. Res. J. **45**(2), 238–273 (2008)

17. Cummins, J.: Age on arrival and immigrant second language learning in Canada: a reassessment. Appl. Linguis. **2**(2), 132–149 (1981)
18. Baker, J.A.: Contributions of teacher–child relationships to positive school adjustment during elementary school. J. Sch. Psychol. **44**(3), 211–229 (2006)
19. Bornschier, V., Chase-Dunn, C., Rubinson, R.: Cross-national evidence of the effects of foreign investment and aid on economic growth and inequality: a survey of findings and a reanalysis. Am. J. Sociol. **84**(3), 651–683 (1978)
20. Kolb, B.: Functions of the frontal cortex of the rat: a comparative review. Brain Res. Rev. **8**(1), 65–98 (1984)
21. Fan, X., Chen, M.: Parental involvement and students' academic achievement: a meta-analysis. Educ. Psychol. Rev. **13**, 1–22 (2001)
22. Hill, N.E., Tyson, D.F.: Parental involvement in middle school: a meta-analytic assessment of the strategies that promote achievement. Dev. Psychol. **45**(3), 740 (2009)
23. Warschauer, M.: Technology and Social Inclusion: Rethinking the Digital Divide. MIT Press, Cambridge (2004)
24. Kay, R.H.: Evaluating strategies used to incorporate technology into preservice education: a review of the literature. J. Res. Technol. Educ. **38**(4), 383–408 (2006)
25. https://www.kaggle.com/datasets/spscientist/students-performance-in-exams
26. Kumar, A., Tripathi, S.L., Rao, K.S. (eds.): Machine Learning Techniques for VLSI Chip Design. Wiley, London
27. El Morr, C., Jammal, M., Ali-Hassan, H., El-Hallak, W.: Data Preprocessing. In: Machine learning for practical decision making: a multidisciplinary perspective with applications from healthcare, engineering and business analytics, pp. 117–163. Springer, Cham (2022)
28. Alturki, S., Hulpuş, I., Stuckenschmidt, H.: Predicting academic outcomes: a survey from 2007 till 2018. Technol. Knowl. Learn. 1–33 (2022)
29. Emerson, J.W., Green, W.A., Schloerke, B., Crowley, J., Cook, D., Hofmann, H., Wickham, H.: The generalized pairs plot. J. Comput. Graph. Stat. **22**(1), 79–91 (2013)
30. Auvinen, T., Hakulinen, L., Malmi, L.: Increasing students' awareness of their behavior in online learning environments with visualizations and achievement badges. IEEE Trans. Learn. Technol. **8**(3), 261–273 (2015)
31. Kumar, A., Sachdeva, M.S.: Detection of mastitis disease in cow with machine learning classifiers. Israa Univ. J. Appl. Sci. **7**(1), 112–129 (2023)
32. Tsao, Y.P., Yeh, W.Y., Hsu, T.F., Chow, L.H., Chen, W.C., Yang, Y.Y., Cheng, H.M.: Implementing a flipped classroom model in an evidence-based medicine curriculum for pre-clinical medical students: evaluating learning effectiveness through prospective propensity score-matched cohorts. BMC Med. Educ. **22**(1), 1–11 (2022)

The manufacturer's authorised representative in the EU is Springer Nature Customer Service Centre GmbH, Europaplatz 3, 69115 Heidelberg, Germany. If you have any concerns regarding our products, please contact ProductSafety@springernature.com

Printed and bound by CPI Group (UK) Ltd, Croydon, CR0 4YY
01/10/2026
02225960-0007